PARENTING IN A CLIMATE CRISIS

A HANDBOOK FOR TURNING FEAR INTO ACTION

BRIDGET SHIRVELL

WORKMAN PUBLISHING
NEW YORK

For the Super Kitties (C, D, J, and S)
and the generations that come after them

Workman
Workman Publishing
Hachette Book Group, Inc.
1290 Avenue of the Americas
New York, NY 10104
workman.com

Workman is an imprint of Workman Publishing, a division of Hachette
Book Group, Inc. The Workman name and logo are registered trademarks of
Hachette Book Group, Inc.

Design by Maggie Byrd
Cover design by Becky Terhune

The publisher is not responsible for websites (or their content) that are not
owned by the publisher.

Workman books may be purchased in bulk for business, educational,
or promotional use. For information, please contact your local
bookseller or the Hachette Book Group Special Markets Department
at special.markets@hbgusa.com.

Library of Congress Cataloging-in-Publication Data is available.

ISBN 978-1-5235-2897-4

First Edition February 2025

Printed in the USA

10 9 8 7 6 5 4 3 2 1

CONTENTS

Introduction

Should I let the kids go outside?

We're thinking of driving to Maine for a few days to get away from the smoke.

What do I tell the kids?

I'm so scared this is the future.

The words came like waves crashing against the shore in a storm, one rolling in after another. It was late spring 2023, and smoke from wildfires in Canada blanketed the northeastern US, turning the sky eerie colors, sending us to check air quality reports before leaving the house. My phone, my moms chat group, my weekday morning walking group were filled with worries and questions. How do we do this? How do we parent at this moment in time? On a rapidly warming planet? On an Earth in crisis?

By the time my daughter was born, I had been writing about food systems and the environment for nearly a decade. I'd spoken with countless worried farmers about soil erosion and what it means for their ability—and inability—to produce food. Chefs, baristas, and mixologists routinely shared their fears about how the changing climate could impact the ingredients they loved. And while I hadn't yet witnessed the apocalyptic skies or experienced the smoke-filled air from increasingly intense wildfires in my corner of the eastern US, I'd listened to colleagues in California talk about the poor air quality that forced their kids inside and the panicked middle-of-the-night evacuations as flames came over the nearby hills. By the time I decided I wanted a child, I thought I knew the full breadth of the climate-crisis horrors a child could face.

Of course, everything you think you know about the climate crisis changes the moment you're looking at your child. As a friend of mine remarked recently, "It's different when you have skin in the game."

In August 2021, the Intergovernmental Panel on Climate Change (IPCC), the scientific group that monitors and assesses the science of climate change, laid out five scenarios for the world our children and grandchildren will inherit, based in part on how humans respond (or don't) to the current crises.

Scenario 1: The world reaches net-zero greenhouse gas emissions by 2050. In all likelihood, the planet warms by more than 1.5°C/2.7°F, but by 2100 the temperature falls back. *We and our children still see the effects of the climate crisis, including an increased frequency and intensity of heat waves, storms, and other extreme weather.*

Scenario 2: The world is slower to reduce greenhouse gas emissions but eventually reaches net-zero emissions after 2050. The planet warms by an average of 1.8°C/3.2°F by 2100. In this scenario, sea levels rise 30–45 cm/12–17 in by the end of the century. I know that doesn't sound like much. *Still,* *it puts an additional ten million people worldwide at risk for coastal flooding, twice as many as the number of people exposed to extreme heat from scenario 1, and there is an increase in extreme weather events.*

Scenario 3: Every country fulfills its existing climate obligations. The world warms by an average of 2.7°C/4.9°F by 2100. *The Arctic Ocean is ice-free in the summer, there is a significant drop in global food production, and there are more extreme weather events.*

Scenario 4: Global cooperation fails, and countries put their interests—primarily their economic interests—first. *The world warms an average of 3.6°C/6.48°F, and sea levels rise 46–74 cm/ 18–29 in, devastating many coastal communities.*

Scenario 5: This is the worst of the worst. We do nothing to make climate change better, and we continue to make it worse. *The Earth warms by 4.4°C/7.9°F, parts of the planet are wholly unlivable, and millions of people, animals, and plants die from the climate crisis.*

To be a parent at this moment, teetering on the edge of climate disaster, is to never live far from anxiety and grief, real fear, intense anger, and immense sorrow. And always lingering underneath my everyday actions— buying groceries, taking a morning walk, watching my child play outside—is this thought: *How will I answer if she eventually asks me, "How could you have had me, knowing what you knew?"*

Like most parents, more than anything, I want my daughter to grow into a happy, kind, successful being, but how do I help her do that on a planet that is getting hotter? How do I talk to my child about it? How will I help her manage her own fears about our planet? How do I manage mine? How do I empower her? And what skills will she need to live in the future?

When much of the world shut down in 2020, my daughter and I found ourselves back in my hometown in a house I had slowly been renovating for several years. Like many people, I spent much of the early days of the pandemic doomscrolling, baking, and thinking about how to keep my family healthy and physically and mentally safe.

I also spent a lot of time staring at a big old dogwood tree in our front yard. Regal, crooked, beautiful, but weathered, the dogwood created a magical bubble over a corner of our yard and front porch. Its rambling branches extended so that when it bloomed in the warmer months, it was nearly impossible to see who was sitting in the porch swing or the sloping yard just beyond. It became a favorite spot to set out a blanket and while away an afternoon reading or playing. Whenever we passed the dogwood, my daughter, who was just learning to talk, started saying, "Bye, tree, bye, tree," in that excited half shout, half whisper of toddlers, and happily saying, "Hi, tree," while patting its trunk whenever we returned.

In May 2020, when the news was unbearable, that old dogwood seemed to ground me to the world, offering a little reassurance. *You've seen some things, haven't you?* I'd muse while pulling the weeds from the flower bed.

And you're still here. In a world turned upside down, the tree was solid earth; saying hello and goodbye to it on our morning walks became a ritual. Occasionally, my daughter threw in a hug.

As the days passed, I started thinking a lot about resilience: the resilience of that 100-plus-year-old tree, the resilience of my local town, of myself, of my child. *Resilience* is a buzzword that gets thrown around a lot these days, but what is it? What does it look like to be resilient? Why are some people more resilient than others? Can parents help their kids develop resilience? Should they?

Resilience is the ability to engage with a challenge and adapt well to it. The biggest challenge facing children alive today (and the children who come after them in the next ten to thirty years) is the climate crisis. Unlike the Covid-19 pandemic, which took the world by surprise, the climate crisis is something we know is coming—it's already here. So how can parents raise climate-resilient kids? How can we help them adapt well to the climate crisis?

I've been an environmentalist for as long as I can remember. Some of my earliest memories are of running up and down the beach outside my childhood home on the Outer Banks of North Carolina. My dad would take me on evening walks, flashlight in hand, to try to spot the crabs scurrying along the sand; my mom would wake me up early to watch the sun rolling over the horizon and help me spot the dolphins playing in the surf; my brother and sister and I spent countless hours jumping in the waves. Idyllic, yes, but I was also aware from a young age of the precariousness of it all.

Storms are scary on the Outer Banks. We'd sit on the covered deck when summer lightning storms rolled in, the stilted house swaying in the wind. When I was a baby, my parents and I once got stranded on a flooded road. A generous passerby stopped to rescue us. That road later became a bridge, and almost every time we went over the bridge or the forecast called for a big storm, we would recount the story of the flooded road. I can recall listening to my parents talk and worry about the beach outside our home becoming smaller and smaller. And as much as I loved watching the dolphins and the other wildlife, especially the pelicans, which I still find mesmerizing, I also witnessed how human-made pollution was killing them.

The entire stretch of barrier islands where I fell in love with the natural world is now at risk. The beaches my parents worried about years ago are shrinking by more than 14 feet a year in some places. The people who live in the towns that dot the shore are weighing their willingness to pay more in taxes and other fees to save the land. It's a fight many know they will eventually lose. The sad reality is that my daughter will not know the Outer Banks like I did. Much of her childhood and adulthood will be different from mine. The climate crisis divides us in this way. The knowledge and experience I have from my childhood feel insufficient. It's why I have sought new knowledge and skills that can help me help her.

For years I've been having conversations with everyone from oyster farmers working to restore clean waters to psychologists pushing for changes in how medicine addresses climate anxiety. These conversations have been about their work, but also about their kids. About how they parent. How they talk to their children about the climate crisis. The skills they think their children will need. The actions they are taking and how they confront their own climate grief.

Before my daughter was born, I got into the habit of weighing my choices against their effect on the environment. Some, like putting solar panels on my house or purchasing an electric car, were easy enough for me to do; others, like divesting the heat source for my house of fossil fuels, or not flying, I find nearly impossible. Others, like composting, I do routinely, then fall out of the habit and struggle to get back into them. But once I had a child, I started questioning more: not only if I was doing enough, but how to raise my child to carry an awareness of and care for the environment that, as an adult, I feel I am learning all the time.

Parents are in an impossible place. Amid the catastrophic headlines, it can be hard to feel hope. We know the climate crisis is an "us problem," not a "me problem." But while many argue that systemic change is what we need to focus on, I don't believe systemic change and individual action are mutually exclusive. Small actions have the potential to drive systemic change. And empowering our kids to know that they can have an impact as individuals lays the groundwork for raising kids who grow up to feel comfortable taking on the more systemic issues as activists or as professionals or simply as voters. There are things we can do to prepare our kids for all the possible climate-change scenarios while helping them develop the skills they'll need to thrive in whatever the world looks like in the future.

I can't claim this book will provide all the answers, but I believe it offers parts of the solutions to many of the issues facing parents as we raise our children in the climate crisis.

I've organized the book into four parts. In Part 1, I explore the big, hard feelings that can color how and when we talk to our kids about the climate crisis. That includes climate anxiety and grief—both ours and theirs. My goal is to help those of us who feel overwhelmed with feelings like fear or anger or despair get to a place of empowerment and hope.

Today's generation of children and the ones who come after them will have to manage the climate crisis as they grow up. Part 2 looks at how helping children find joy in the natural world can be an antidote to climate grief—and a way to foster a sense of responsibility for protecting the environment.

In Part 3, you'll find tactical advice about taking real steps to live a more ecologically sound and healthy life and about how to build resilience in the face of so much uncertainty. Part 4 looks at how individual actions can help to create lasting systemic change.

I believe that we can mitigate the worst of the climate crisis, that we can adapt well to what we cannot change, and that in the process we can create a better, fairer, more sustainable world.

Our kids' lives and their world will be different, but that doesn't mean it can't be beautiful.

Part 1 `Feel`

I'm assuming that the majority of you who picked up this book did so because you have some big feelings around raising kids in a climate crisis. You may be afraid. Or angry. Or really worried. Or simply unsure where to start with helping your kid understand this big issue that maybe you don't even feel like you understand. You're fielding questions from your child such as "Is the planet dying?" and "Why aren't people doing more?"—and you wonder the same things yourself.

I set out to write a book that is empowering and focuses on resilience and hope, but to get there, first we have to learn to recognize feelings of climate grief and anxiety—in both ourselves and our kids. Understanding these emotions is a necessary skill for anyone hoping to make a difference and find resilience in the face of so much uncertainty.

▰▰▰1▰ Talk

The beautiful everyday chaos of parenthood fills my days. There is preschool pickup and drop-off and work to squeeze in between. There are building blocks and other toys to pick up and walks to pet the goats at the farm near our house. There is mac 'n' cheese to make and more work to squish in while my child spills the building blocks all over the floor yet again. And all the while, never too far below the surface of my daily mothering, I feel the thrum of worry, fear, sadness, and anger over the warming of the planet that humans have caused (and are continuing to cause)—simmering feelings that threaten to boil over and steal my attention.

My child is too young to fully grasp why there isn't snow in July, let alone the broad scope of the climate crisis. But she will eventually. And while it's tempting to hide the climate crisis from our children for as long as we can—after all, we want the world to be perfect for them—they need to know and they deserve to know.

My child and yours will have to navigate the climate crisis for much of their lives. Many of them already are. In late spring 2023, when smoke from wildfires in Canada engulfed much of the northeastern US, including our coastal town, for a week, we had our first taste of that life. When my child asked to go outside after school, and I said, "Yes, for a few minutes, but only if you wear a mask," I felt a stab of horror at how practical it sounded. My child, so used to masks from the pandemic, looked at me and asked, "Because the air can make me sick again?" Yes, I said, and she accepted that, but eventually, I know she'll ask why, if we can fix it, and how.

As our children grow, they'll face days of poor air quality and high temperatures, forcing school closures, event cancellations, and more. They'll experience (or see on the news) severe hurricanes, wildfires, heat waves, and other natural disasters we can't yet fathom. The climate crisis will, eventually, inform their every choice, from what they eat to how they travel to where they live. They'll need to understand how we got here and how to evaluate information about climate change. And amid all that, we don't want our children to live with constant anxiety. We want them to still be kids.

"My oldest wants to know why more people aren't doing more, which then feeds into her questioning of why she cares and others don't," says Chantal Alison-Konteh. The New York City mom and stepmom of four children ranging in age from four to nine is an educator and tells me she's barely started to talk to them about the larger issue of global warming and fossil fuel emissions. But recently her nine-year-old and her eight-year-old, whom she homeschools, did a project on plastics in the ocean, which led to more questions—specifically, trying to understand why more isn't being done.

So, how does Alison-Konteh broach the topic? How do climatologists? How do teachers? Is eight too young? When is the *right age* to talk about the subject? How should parents have the talk? And how do parents find the line between informing and alarming our children?

Talking about global warming doesn't have to be a formal, sit-down conversation, and it's definitely not a one-and-done discussion. And depending on the age of your child, how much and what details you share will vary.

Climate Crisis 101

Before you bring up the subject, it helps to brush up on some basics. Conversations about climate change can feel intimidating, especially if you don't have a science background, but understanding the terms used and basic science can empower us to talk about it.

Weather is what happens each day, including how hot or cold it is, whether it's cloudy or sunny, raining or snowing.

Climate refers to the average weather in one place over a set time. This "one place" can be a particular area, such as a town, region, or country—or it can be the Earth as a whole.

Climate change is just what it sounds like: changes in the usual weather patterns, such as how much snow a place gets in a given year or a change in the Earth's average temperature. Just like with our growing kids, it can be hard to see the changes in the day-to-day until all of a sudden you realize at some point that you've tied their shoes for the last time, or that it's been five winters since you needed to use the snow shovel.

How do we know that the climate is changing?
Because scientists have been studying and document-
ing these changes for more than 100 years. And while
systemic scientific assessment of climate change didn't
begin until the 1970s, the living things on our planet leave
a record of the Earth's changing climate that goes back
centuries. Scientists routinely use the insides of trees, ice
cores, and lake and ocean sediment to find evidence that
tells them what the temperature was during a specific
year.

So what is causing the Earth's climate to change?
We are. Yes, many things can cause the climate to change,
such as the Earth's distance from the sun, or events like
volcanic eruptions. Still, there is no doubt that humans
are changing the climate. Studies show that emissions of
greenhouse gases from human activity have led to at least
1.1°C/1.9°F of warming since the late 1800s and early
1900s. It may seem like a small number, but think of it in
terms of your child's body temperature—an increase of
1.1°C/1.9°F means your child has a fever. When it comes
to the planet's fever, every degree matters. If you're a
longtime gardener, you've probably already noticed
how the warming planet has affected what you grow.
In 2023, the US Department of Agriculture updated its
Plant Hardiness Zones, and vast swaths of the US are
now in warmer zones. Scientists tell us that not only has
the number of hurricanes been increasing but so, too,
has the number that become Category 4 and 5 storms.
The temperature can even affect how many male or
female turtles are born, because most turtles experience
temperature-dependent sex determination—warmer temperatures mean more females. And at 2°C/3.6°F of

WE'VE KNOWN ABOUT HUMAN-FUELED GLOBAL WARMING SINCE 1856

You probably never learned about Eunice Newton Foote in school, but you should have. We all should have. Foote is considered one of the first climatologists. In 1856, the American scientist conducted experiments using an air pump, glass cylinders, and four thermometers. She was testing the impact of carbon dioxide (known then as carbonic acidic gas) on standard air. She found that when placed in the sun, the glass cylinder with carbon dioxide trapped more heat and stayed hot longer; this was the start of what we think of as the greenhouse effect today.

"An atmosphere of that gas would give to our earth a high temperature; and if as some suppose, at one period of its history the air had mixed with it a larger proportion than at present, an increased temperature . . . must have necessarily resulted," Foote wrote in her paper "Circumstances Affecting the Heat of Sun's Rays," which was presented at a meeting of the American Association for the Advancement of Science and later published.

Foote didn't know the impacts of an increase in the Earth's temperature. Still, she correctly theorized that the Earth's temperature would increase because of a rise in greenhouse gases.

warming, we'd see five times the number of floods, storms, and heat waves, according to the IPCC.

What are greenhouse gases? Carbon dioxide, methane, ozone, nitrous oxide, chlorofluorocarbons, and water vapor are known as greenhouse gases. The name refers to their ability to trap heat in our atmosphere (like a greenhouse) and increase the Earth's temperature. According to the US Environmental Protection Agency (EPA), carbon dioxide accounts for nearly 80 percent of all human-related greenhouse gases. You'll often hear people use *carbon* as a blanket term to include all human-made emissions.

How do humans cause climate change? In the 1950s and early 1960s, thanks to more precise measurements and computer models, scientists were making the connection that the increase in greenhouse gases in the atmosphere was the result of burning fossil fuels. In 1965, climate scientists sent then US president Lyndon B. Johnson a report called *Restoring the Quality of Our Environment* with this warning:

> Through his worldwide industrial civilization, Man is unwittingly conducting a vast geophysical experiment. Within a few generations he is burning the fossil fuels that slowly accumulated in the Earth over the past 500 million years. . . . The climatic changes that may be produced by the increased CO_2 content could be deleterious from the point of view of human beings. The possibilities of deliberately bringing about countervailing climatic changes therefore need to be thoroughly explored.

In the nearly sixty years since that alarming warning, attempts at curtailing fossil fuels have met steady resistance, and carbon emissions from all sectors have increased by more than 200 percent. Documents unearthed by researchers, journalists, and other concerned citizens show that energy companies have continued to invest in fossil fuels while using marketing campaigns to delay climate action.

Starting the Conversation and Keeping It Going

How and what you share about the climate crisis will depend largely on the ages of your kids and, to some extent, their personalities.

"The balance is always to share what's developmentally appropriate for your kids without scaring them," says Dr. Abigail Gewirtz, a child psychologist and author of the book *When the World Feels Like a Scary Place*. No matter your kid's age or temperament, the essential things to stress are that the climate crisis isn't their fault, that people worldwide are working toward solutions, and that there are steps we can all take to help.

Ages 4 and Under: Loving Nature

My daughter is a nature kid. She delights in the outdoors, as do most young kids. Although babies, toddlers, and preschoolers are too young to grasp the concept of the climate crisis, we can foster their joy in the outside world.

Continued on page 11

A Timeline of What— and When—Fossil Fuel Companies Knew About the Climate Crisis

- **1959:** Scientist Edward Teller gives a speech at the Energy and Man symposium celebrating the 100th anniversary of the global oil industry. Teller warns, "Whenever you burn conventional fuel, you create carbon dioxide. . . . The carbon dioxide is invisible, it is transparent, you can't smell it, it is not dangerous to health, so why should one worry about it? Carbon dioxide has a strange property. It transmits visible light but it absorbs the infrared radiation which is emitted from the earth. Its presence in the atmosphere causes a greenhouse effect. It has been calculated that a temperature rise corresponding to a 10 per cent increase in carbon dioxide will be sufficient to melt the icecap and submerge New York. All the coastal cities would be covered, and since a considerable percentage of the human race lives in coastal regions, I think that this chemical contamination is more serious than most people tend to believe."

- **1965:** Frank Ikard, president of the American Petroleum Institute (API), briefs a group of fossil fuel executives on *Restoring the Quality of Our Environment*, a report by President Johnson's Science Advisory Committee, saying, "The substance of the report is that there is still time to save the world's peoples from the catastrophic consequences of pollution, but time is running out. . . . One of the most important predictions of the report is that carbon dioxide is being added to the Earth's atmosphere by the burning of coal, oil, and natural gas at such a rate that by the year 2000 the heat balance will be so modified as possibly to cause marked changes in climate."

- **1971:** European-based oil company Total becomes aware of the potential for catastrophic global warming. By the late 1980s, Total is attempting to cast doubt on global warming by disputing the science.

- **1979–83:** The API and representatives of the major fossil fuel companies form the CO_2 and Climate Task Force to monitor and discuss the latest climate research.

- **1979:** Exxon studies ways to avoid global warming, including transitioning to renewable energy.

- **1980:** The API's CO_2 and Climate Task Force, which has changed its name to the Climate and Energy Task Force, receives a briefing on the latest climate science from John Laurmann, a scientist from Stanford University. Laurmann warns that continued use of fossil fuels will lead to catastrophic effects from global warming by the 2060s.

- **1980:** Despite Laurmann's warning, the API tells governments there will be no negative consequences from tripling worldwide coal production.

- **1981:** Exxon manager Roger Cohen sends an internal memo saying that it is possible fossil fuel combustion could "produce effects which will indeed be catastrophic (at least for a substantial fraction of the earth's population)."

- **1982:** Exxon internally circulates a forty-page report on climate change, predicting (fairly accurately, as it turns out) the amount of global warming we've experienced since then.

- **1986:** Shell circulates a nearly 100-page report predicting that fossil fuel–caused global warming could mean "destructive floods" and more.

- **1989:** Exxon and other fossil fuel companies begin a coordinated global effort to deny climate science, lobby lawmakers to block clean energy legislation, and more.

- **1990s:** Throughout the 1990s, fossil fuel companies dispute climate science and confuse the general public with ad campaigns that include slogans like "Doomsday is cancelled. Again."

- **1998:** A memo from the API details the plan for climate deception, claiming that "victory will be achieved when . . . average citizens 'understand' uncertainties in climate science."

- **2001:** ExxonMobil continues to work to undermine climate science, seeking to influence how the Bush administration reports on climate science to Congress.

- **2004–06:** BP popularizes the term *personal carbon footprint* as fossil fuel companies work to put the responsibility of climate change on individuals.

While most fossil fuel companies no longer outright deny climate science, their long history of hindering progress continues in other ways. They deflect their own responsibility and put the onus on individuals, they create doubt around possible solutions, and they even outright work against them, such as car companies that sue over efficiency requirements and companies that lobby lawmakers to make it illegal for cities to ban gas stoves.

Continued from page 7

Lean in to your young child's love of nature and the curiosity that love can inspire. This is what Dr. Erica Smithwick, the associate director of the Institute of Energy and the Environment at Penn State University, and Dr. Rosimar Rios-Berrios, a scientist at the National Center for Atmospheric Research in Boulder, Colorado, recommend. Both are members of Science Moms, a nonpartisan group launched by the Potential Energy Coalition in 2021 to help educate caregivers about the science behind climate change.

If you don't already, incorporate outdoor time into your daily routine. Set aside five, ten, or fifteen minutes at the beginning or end of the day to take walks around the neighborhood or your backyard. As your child finds their footing, let them lead, throw stones into the water, touch flowers, plants, and trees, and observe the bees and birds. Help open their eyes to the natural world by pointing out how elements interact with one another (like the bees pollinating the flowers) and how nature can change (like how the grass changes color when it's hot and dry).

This time outside sets the foundation for your child's love and respect for nature.

"When we go berry picking, I teach them what my grandmother taught me," says Yatibey Evans, a mom of four in Alaska. "You don't pick all the berries off a bush. You save some to grow for the next year and for the animals that might come and eat the berries . . . and you say thank you to the plant for providing this food for you and your family."

At this early age, it's about letting kids discover and fall in love with the outdoors while learning simple actions to take care of the planet.

"We should be talking about the environment with our kids," says Rios-Berrios. "Explaining [to them why] we do things like turn off the lights, and encouraging them to spend time outdoors—all those things are excellent starting points."

And when things get a little scary—say, when the air quality drops and you need your child to wear a mask—you can find age-appropriate ways to explain it to them.

Young children's brains don't operate in the abstract, so concrete examples are necessary.

"Four-year-olds are pretty black and white. They don't understand abstract concepts," Gewirtz says when I ask her how to talk to my own child about wearing a mask during the wildfires. She recommends being as specific as possible

HELPING YOUNGER KIDS THINK ABOUT WEATHER

- Ask your kids to think about how the weather affects the clothes they wear on any given day.

- Walk around your neighborhood with your kids on the next rainy day. Explain to them where water comes from and where it goes. Build a rain barrel to conserve water from your gutters. Use that rainwater to water a vegetable garden.

- If your child wants to know when it's going to snow, talk about what temperature it needs to be to snow and have them track the daily temperature. As they get older, explain the mechanics of how snow happens. Have them look back at the average daily temperature they've been tracking and use it to talk about climate.

about the threat. "You don't want kids to think that everything outside can make them sick, so [focus on] the smoke."

It's important to relate it to concepts they are already familiar with, Gewirtz adds. "For example, if you have a fireplace, you can explain how the smoke goes up the chimney instead of coming into the room. You can explain that sometimes when there are big fires that are outside, even ones that are really far away, the planet's chimney doesn't work so well, so we have to find another way to not breathe in the smoke while we're outside."

Ages 5–8: A Growing Understanding

"The first time we had an explicit conversation about the climate crisis was when [my daughter] was entering first grade," says Dr. Tamara Yakaboski, an educator and climate-change coach in Colorado. "I took her to a climate strike, and she made a little sign that said something like 'Love your mother,' and it had a little Earth on it."

By the time kids are in elementary school, they have an understanding of seasons and weather, and chances are they've seen or heard enough on the news, in school, and from their peers to have a sense that something is up with the Earth's climate.

Let their questions about the world and the weather guide your chats. Most kids at this age will need help to grasp the difference between weather and climate, but you can start setting the stage for how they understand it.

Briana Warner is a mom of two and the CEO of Atlantic Sea Farms, a kelp company based in Maine. Warner's two boys know about climate change because it's the focus of her work, but her kids may be in the minority. According to a 2022 report by the Siena College Research Institute, only about half of parents (49 percent) said they had talked to their children about climate change. That's despite the fact that the same report found that nearly eight in ten Americans agree that climate change is a "very" or "somewhat" serious problem.

Warner's first grader asked her to come to his class to talk about what she does. When she asked, "Who here knows about climate change?," only two kids raised their hands, and one of them was hers.

"I didn't call on him, because I knew what he was going to say, so I called on the other kid, and he said it's like a big warm blanket wrapping the Earth and making it warm. I'm like, 'You know what? Yes. What you said, that's totally right.' His parents clearly met him where he was."

Kids at this age shouldn't feel that it's up to them alone to solve the climate crisis, but helping them see that they can make a difference will empower them.

Kids are naturally curious, Yakaboski says, and we can take advantage of that natural curiosity. "I've always tried to answer [their questions] in ways that are not fear based, but [that] this just is what it is," she adds. When her daughter was in second grade, the students had to pick a passion project. Her child decided to do one on trash pickup because she was naturally curious about why people litter.

You can also lean in to their interests. "I've been trying to find different ways to teach [my] girls about what's happening in the world," says Alison-Konteh. "We watched a documentary on the ocean, did an ocean study, and now my daughters are trying to convince us not to use plastic anymore. We all have stainless steel water bottles, and

HOW TO SEPARATE FACT FROM FICTION

There is so much misinformation and conflicting information about the climate crisis that it's essential to evaluate the source of the climate information we're reading, hearing, and seeing. Helping our kids do the same will make them better informed and teach them to think critically. I like the SIFT method, developed by digital literacy expert Mike Caulfield, for determining whether information is reliable. Here are the four steps of this method:

1. *Stop and think*: Before reacting to information, take a moment to pause and consider what you're reading, hearing, or watching.

2. *Investigate*: Where is the information coming from? In other words, who created it, and

we're big into recycling. It really matters to them, and I'm learning with them. It's definitely been making me more conscious of my plastic usage and waste."

Ages 9–13: Facing Fears, Finding Solutions

"He's asking questions that are heartbreaking," says Isabel González Whitaker, who works with Moms Clean Air Force, a nonprofit, grassroots environmental advocacy what is that organization's or person's background? Do you understand what, if any, the author's political, ideological, institutional, or personal biases are? Is there contact information for the source? What's the purpose of the information? Is it fact or opinion? Is the author trying to make an argument?

3. *Find other coverage*: Is what they are telling you broadly accepted? Are other people and organizations saying the same things? Is it supported by evidence?

4. *Trace*: What is the date of the information, and when was it last updated? Follow the claims, quotes, and images to their origin. It's also important to ask yourself who is telling the information and who is benefiting from the information.

group, and is mom to a ten-year-old. During the summer of 2023, with wildfire smoke in the air, images of the devastation in Maui on social media and in the news, and extreme heat elsewhere in the world, her son expressed some real climate anxiety. "His questions were really doomsday. I try to be honest. I say, 'This is why I do the work that I'm doing. This is why we try to recycle to the best of our ability. Why we try not to buy one-off plastics.' You've got to tell them the truth, and you've got to tell them in ways they can understand. And you can't lose hope."

If you haven't already, this is a good time to start explaining the carbon cycle on a microscale. You can explain how plants make food and grow by taking in carbon dioxide, water, and sunlight. From there, you can start to talk about the carbon cycle on a global scale. While carbon dioxide helps keep our world warm by trapping heat in the atmosphere, too much of it can change the Earth's climate, and over the past 100-plus years, we've been producing too much of it.

By the time they're this age, you can start playing to your kid's personal strengths. They are at the age when they are likely more optimistic about change than teens are. Lean in to that optimism and help them organize a letter-writing campaign to Congress, or see if they and you can participate in a beach, river, or park cleanup, or challenge them to find single-use items around the house that you could swap for reusable items.

"You're sharing about the issues, such as how climate change contributes to forest fires, but you're also sharing what we can do about them," Gewirtz says.

Continue to make time for your kids to be outside every day. For some, it's sitting outside on a deck, porch, or even under a tree and debrief about the day, or if your child is fascinated by birds, take walks to see how many different kinds of birds you can find.

If your kid loves research, go back to that rain barrel (see the box on page 13) and have them track how much water they can get out of it in any given month, and then look up what the average rainfall was twenty years ago, thirty years ago, and so on.

Ages 14–17: Values and Voice

By this age your kid might be teaching you things about climate change. Let them take the lead in conversations about the climate crisis. Listen to the information they want to share (and help them find accurate sources). Be open to changes they want to make in the household, such as setting up a compost bin or becoming a vegetarian. Even if you don't want to give up meat, you can put them in charge of the dinner menu for a night or make compromises about how frequently you eat meat or from what sources. The important thing is that they feel like they're making an impact in their home.

Help them identify the things they love and that they're good at, and help them find ways to use these interests and talents to tackle parts of the climate crisis. Your super-social teen could organize a fundraiser for a local climate project. Your varsity lacrosse player could suggest ways to reduce waste at home games. Your research-obsessed kid could attend some town hall meetings and learn about local issues. Empower them by letting them independently find resolutions to certain parts of the climate crisis. Listen, then ask how they feel and what they need help or support with.

This is also an age when you'll want to keep an eye on how the climate crisis might be affecting their mental health. It's easy for us all to feel overwhelmed by the climate crisis, but this is especially true for teens, who often feel like they can't control what is happening. Encourage them to practice self-care: learning meditation or relaxation techniques, making time to journal or exercise, and, of course, spending time outdoors. Things look brighter in the fresh air.

"Being outside is hugely important," says Clare Flaherty, a seventeen-year-old who lives in Narragansett, Rhode Island, and an ambassador at the Climate Initiative. "It's how my parents raised me and my siblings. We moved around a lot, but we were always outside no matter where we [lived], and being outside still plays a role in my day-to-day life. I feel like my generation has a handle on the importance of well-being and we prioritize it, and for me the connection with the outdoors and nature is important. It's how I ground myself. Even if it's just sitting on the beach for five minutes, it gives me a sense of contentment."

Focusing on the specific ways they can make a difference can keep their hard feelings from tipping into despair. Start by being vocal about how you're doing your part and how caring about climate change is part of a value system that you want to pass down to them. Encourage them to live mindfully in the world and be aware of the impact that our choices have on the planet.

"I'm a vegetarian and I'm going to raise my whole family vegetarian, and we are going to buy the eco-friendly car, not the fast car that my teen would love," Smithwick says, explaining how she uses daily life decisions to talk about climate actions with her children.

Brooke Petry, a South Philadelphia mom to a teenager and a field organizer with the environmental group Moms Clean Air Force, uses the challenge of living in eco-friendly ways to talk about not getting discouraged, and understanding issues of privilege.

"I always try to communicate about how nobody is perfect and it's okay to not be perfect. And not for us to really judge. I never want it to be like, *We're doing this and everyone who doesn't doesn't care about anything*, because there's a lot of layers, right? Economics, access, privilege—it's all in there."

You can also talk to your children about what you care about. "We don't have a lot of conversations now except [about how I am] so cringe," says Evans about her fifteen-year-old. "[But then] I had a conversation with him about how I was advocating for people to oppose land relinquishment [which would allow other uses for the land, including the possibility of mining it], and he said that would be horrible. For him to have that emotional reaction, I felt all the things that I've been teaching him as he's been growing up have kind of made a difference."

No matter what stage of parenting you're in, remember that just as the climate crisis isn't a onetime thing, neither is the conversation. Be prepared for your kids to continue to have questions, fears, and other emotional responses. You'll need to repeatedly follow up on discussions about the climate crisis, and you need others on your team.

As my daughter grows up, she's not going to share everything with me, and that's okay, but I want her to have people she can turn to when she needs them. Helping our kids deal with the climate crisis means helping them build a support system of trusted friends and adults whom they can speak with and lean on when necessary.

These conversations should be hopeful. As easy as it is to get sucked into talking about the challenges of the climate crisis, we also need to discuss the positive changes. Remind your kids that people around the world, including adults and other kids, are working on climate-crisis solutions and that they can help.

If You're Not Talking to Other Parents About Climate Change, It's Time to Start

Talking with other adults about climate change is one of the best things we can do to tackle it, according to Science Moms and Parents for Future UK. Honestly, the first time I heard that idea, I didn't love it. Still don't. Talking feels too much like inaction to me. And as a friend recently said, "I feel like most of my friends and family are so sick of hearing me talk about it, they tune me out." I get that—but I do it anyway. Sometimes I find it hard not to scream at people, *How can you be so calm, how can you quietly go about buying your single-use plastics, driving your gas-powered cars, voting the same people into power who have always been in power? Don't you realize the world is on fire?*

But even if your friends and family aren't talking about the climate crisis, they're likely thinking about it. Talking brings those thoughts into the light, which is the first step in taking action.

"For my child to see me talking about climate change, talking to other parents about climate change . . . I think one of the most important [things] we bring to the table as Science Moms is that our children see us being activists," says Rios-Berrios.

Seven in ten Americans think the climate crisis is a danger today, yet according to a 2019 study by Yale and George Mason universities, fewer than four in ten Americans (37 percent) "occasionally" or "often" talk about global warming with their friends and family.

"Parents for Future's work is very much focused on systemic change, not individual change," says cofounder Rowan Ryrie. "So this is about trying to make sure that we as a community of parents can put pressure on the systems that need to change and on the power holders in society, not on individuals changing individual behavior. Parents, particularly parents of young children, are one of the most socially connected groups in society. We are networked into other communities and other kinds of groups across our community in ways that a lot of other networks are not. Getting parents to talk about climate in those interconnected spaces can be impactful."

We trust our friends and family, and when we talk about important issues, like the climate crisis, we help spread knowledge and build the foundation to take action together. "Discussion with others in one's close social network . . . appears to be an important route by which people learn key facts about an issue, such as the scientific consensus," a 2019 report found. "Without discussing global warming, people may never learn important facts about climate change or that close friends and family care about the issue."

Everyone I spoke to for this book agreed on one thing: When talking with friends, family, and casual acquaintances about climate change, we must make it personal.

Science Moms suggests talking about feelings before facts. If you're feeling worried about climate change, say it out loud. Or say you're frustrated when your kids' school doesn't compost.

You can also ask questions. You likely know people who are skeptical of climate change or who don't see how climate change applies to them or who don't feel the need for urgent action. You can also find indirect ways to get to the issue. Poll your friends for good vegetarian dinner ideas, and share your reasons for eating more plant-based foods (yes, this is a climate action). Mention to people that you recently added solar panels to your house. Or share why you decided to take a summer trip that was within driving distance instead of flying.

"We need everybody engaged, and we need people to start talking about what changes they can make in their own lives," says Smithwick. "It is different for everybody. We have to acknowledge that most people are not in a position to make every behavioral change. They don't have the money to buy a Tesla. They don't have the ability to bike to work. So we have to meet them where we can. One thing I've started to be more proactive about is explaining choices, like, 'By the way, did you know you can switch your electric distribution to renewable energy at no cost to you? And it only takes a phone call, and oh, by the way, I found a better rate at this company.' Then all of a sudden people are talking about energy transitions and doing it in a neighborly way. We have to start those conversations. Otherwise, the change doesn't happen."

Whether you ski, love to watch soccer or football, travel, or drink pumpkin spice lattes, you know it: Everyone cares about something that the climate crisis is putting at risk.

"There's lots of different ways to take action, but what we try to do with Parents for Future is lower some of the barriers to stepping in and show people that just talking about it is often a really good entry level. Then people can go on to do more once they get more confident," says Ryrie.

There is a growing misconception that it's too late to do anything about the climate. When people feel doomed, they are less likely to act—but our children need us to act. As climate activist Greta Thunberg said, "When we start to act, hope is everywhere." Research backs this up. People who feel constructive hope are more likely to act and support climate policies, according to the Yale and George Mason report. Making the large-scale social changes necessary to avoid the worst impacts of the climate crisis requires social support. We can cultivate that support by sharing our feelings and talking about our actions.

▮2▮ Grief

Watching elephants toss up dirt as their own brand of sunscreen in Tanzania's Tarangire National Park, I found it impossible not to smile. They looked like big goofy dogs. I wasn't a mom then, but as I watched those silly, majestic creatures, I pledged that when I did have kids, I'd find a way to bring them to see these elephants in the wild. Now I'm torn. I want my child to see those elephants and many other things. I want her to fall in love with the world, but there are times when I wonder if she's better off never seeing firsthand what we're losing.

In 2021, the International Union for Conservation of Nature (IUCN) listed the African forest elephant as critically endangered, reporting that their population had decreased by more than 86 percent over thirty-one years. The African savanna elephants like the ones I saw in Tanzania are also on IUCN's endangered list—their population dropped by more than 60 percent over fifty years. Unless something radical happens, all elephants will likely be extinct within our lifetimes. In addition to being susceptible to high temperatures, elephants face threats such as poaching, habitat loss, and the potential loss of fresh water, all of which the climate crisis will exacerbate. And they are far from the only creatures we are losing. Since 1970, the world's wildlife populations dropped by more than 69 percent, according to the Living Planet Report 2020.

Of course, it's not just animals thousands of miles away that we're losing, but also parts of our own childhoods that our children will never get to experience. It's rare these days for us to spot fireflies in our backyard, yet I remember watching them outside my window as a child. My daughter has soccer practice indoors some days, not because of rain but because of smoke. We planted apple trees in our yard several years ago, and I wonder if they will be able to survive climate threats. Maybe. But some of the streets in our coastal town, ones we walk down every day, will be lost to rising waters.

A friend recently mentioned wanting to take her five-year-old to see the ancient sequoias before they're gone. The US Forest Service estimates we've lost up to a fifth of those beautiful giant trees because of recent wildfires. Another parent mentioned her child asking why there hasn't been enough snow to use their skis. How do we help our children manage their climate grief without passing on our own?

"My family went blueberry picking all throughout my childhood, but we can't go there anymore, because it's been devastated by this mine, and it's been devastating for our people in that region," says Yatibey Evans. Evans is Ahtna Athabascan from Mentasta, Alaska, and says she feels a deep connection to the land in part because of the values of her grandmother who, for most of her childhood lived a life following animals and plants for sustenance. "I believe our connection to the land is really what drives us, keeps us alive, sustains us, and without that connection, I feel like a big part of me would die."

There are many forms of climate grief. There's the immediate grief and trauma of tangible loss, such as losing a home to a wildfire, or the heightened anxiety of a particularly bad tornado season. Then there are the harder-to-define feelings of both anxiety and loss over what we're losing but haven't yet lost. In 2007, Australian philosopher Glenn Albrecht gave those feelings of climate distress a name—*solastalgia*—a haunting, beautiful word for the anxiety, guilt, grief, and anger so many of us, including our kids, are feeling about the climate crisis. *Imminania* is another term, one that describes a feeling of sorrow for the future.

"It's heartbreaking. You have to balance so many emotions—grief, heartbreak, and then anger," says Alexina Cather. "I'm angry because I feel like our generation didn't make certain choices and now we're having to suffer, and we're going to see our kids suffer even more."

Cather, who has worked in food policy for the Hunter College New York City Food Policy Center, the James Beard Foundation, and Wellness in the Schools, is a mom of three, including an infant.

"With each child I've gotten more anxious about the climate," Cather tells me. "I think every kid just kind of opens your heart a little bit more, but by the time I had my youngest, I had been working on food and climate for the last decade. So I'm much more hyperaware."

She lives in Vermont with her family. During the devastating summer floods of 2023, her husband, a volunteer firefighter, was out helping the community while Cather sat up in the middle of the night with her three boys.

"I just remember being upstairs, like, *What am I going to do? Well, I have two hands, but how am I going to keep all these kids safe if there's a real flood, if the river breaks?* We live very close to the river and we were really lucky, but it was very real. I remember sitting up that night and feeling so much anxiety. Then after that crisis there were floods in other places, and fires, and it feels very—I think the word is *heavy*, but it's also almost oppressive. Like it's hard to breathe sometimes as a parent."

While there's no single accepted definition of *climate grief*, Oxford Languages defines *eco-anxiety* as "extreme worry about current and future harm to the environment caused by human activity and climate change."

When you're a parent, climate grief can feel like anxiety and sorrow mixed together.

According to a 2023 Yale survey, anxiety is rising in the US over the climate crisis. Sixty-five percent of Americans surveyed said they were at least "somewhat" worried about the climate, down slightly from 70 percent in 2021, but up from 62 percent in 2018 and 49 percent in 2010. The percentage of those who described themselves as "very" worried was 35 percent, up from 21 percent in 2018.

Globally, among young people ages sixteen to twenty-five, there is even more anxiety and anger. According to a 2021 survey by *The Lancet Planetary Health*, nearly 60 percent of people ages sixteen to twenty-five were very or extremely worried about climate change, 40 percent were hesitant to have children because of the climate crisis, and almost 50 percent said anxiety or distress about the climate crisis was affecting their daily lives.

"I've always been pretty aware of the climate crisis, but I wouldn't say it impacted my decision to have kids," says New York City mom Chantal Alison-Konteh. "I was in high school when Hurricane Katrina happened, and I remember it being really scary for me. I'm someone who feels the pain of the world deeply, and when Hurricane Katrina happened and I heard all those stories, I kept wondering, *How could this happen?* That's when I really started to learn about the climate crisis, but I didn't think it was important to me until I had kids, because it's important to them."

The American Psychiatric Association (APA), which is one of the largest professional psychiatric organizations in the world, considers the climate crisis a growing threat to mental health.

"Mental health professionals help people face reality, because we know living in denial can ruin a person's life," said psychiatrist Lise van Susteren in a 2017 APA report on mental health and our changing climate. "As the climate crisis unfolds, we see people whose anger, anxiety, and depression, caused by the shortcomings of a previous generation, prevent them from leading productive lives themselves."

Multiple studies are already showing the effects of the climate crisis on our health. According to the *Lancet* Countdown on health and climate change, children are more impacted by disasters than adults and are more likely to have continued trauma-related symptoms after a disaster. Other studies link extreme temperatures with an increase in hospital admissions for mood and behavioral disorders and an increased risk of suicide.

Yet mental health practitioners and systems are struggling with their role in the climate crisis.

"We're absolutely not prepared," says Dr. Tamara Yakaboski, who hosted Embodied Climate Conversations with the Somatic Nature Therapy Institute. "Part of that is just that here in the US, climate is so politicized that there's a lot of debate among medical providers, some of whom will say, 'We can't really talk about it,' or therapists will wonder if asking someone directly about climate grief is leading."

We are starting to see that change, especially in the hands of parents. In the UK, Charly Cox founded Climate Change Coaches in 2016 after the birth of her daughter. The organization seeks to help people make sense of the emotional and practical impact of climate change so they can take meaningful action. Similar organizations and grassroots groups, such as Embodied Climate Conversations, are popping up around the world.

The Embodied Climate Conversations that Yakaboski hosted took place online and were made up mostly of therapists whose patients are increasingly bringing up concerns about the climate crisis.

"It's a very different kind of grief," says Yakaboski. "We talk about how we help people, how we sit in real shitty emotions and learn how to tolerate them so that we can do resilience building."

Yakaboski also started a group of what she calls the backyard climate feminists. She describes the group as made up of educated, kick-ass women, many of them mothers, all of them concerned about the climate crisis, who meet in her backyard to talk through their emotions around the climate crisis and what actions they can take.

"Finding community and validation helps build resiliency," says Yakaboski. "It's helping people find what they love enough to save, so we can avoid the paralysis of what to do."

It's the work that is part of living with climate grief that has the potential to alter our world for the better.

The Signs of Climate Grief

One summer, my daughter and I spent a month on the beach in rural Nova Scotia. The beach felt wild and untamed, and many days we were the only ones on it. It reminded me so much of the Outer Banks of my childhood that some days it hurt. The place that first made me an environmentalist doesn't exist anymore. Lost to development and over-tourism, it will eventually be claimed by the sea. Yet despite my sadness, the memories also brought me joy, mixing my past with my present as I watched my own child connecting with nature on this new (to us) beach hundreds of miles away.

It's one thing to know the definition of climate grief, but what does it look like? Is it similar to other kinds of grief? We're a bit grief averse in the US. We don't like to deal with it, and because of that, I think we like to create patterns out of our grief, or at least try to find them.

Climate grief can look like a lot of different things:

Denial: Even those who believe in human-fueled climate change likely feel some denial about how serious the problem is or that it's not something to worry about, because it will get figured out before it gets too scary.

Anger: You may be angry over the government's lack of action toward solutions, or angry at other people's lack of interest or action. You may also feel anger that the system feels stacked against us. All of this is normal. Anger can be a powerful motivating force! But it can also be paralyzing.

Depression: It's easy to get depressed and overwhelmed about the climate crisis. Depression can manifest as feeling numb to the situation or feeling so much sorrow and helplessness that it feels impossible to take any action.

Climate grief is simply something we all must live with now. Some days it can be overwhelming, especially when we're reading one awful headline after another.

"The only way you're going to be okay in this moment is if you accept the darkness in with the light," says Michele Bigley, a writing instructor and travel journalist who focuses on environmentally conscious tourism. Bigley sees a lot of climate-related mental health issues in her students and in her own teens. "We have to accept the reality that we're not going to fix it. We're going to learn to live with it. We're going to find skills, we're going to adapt to it, and we're going to work really hard to make changes, but we're not going to fix it. And that's a big lesson. We don't have to bear the responsibility of trying to fix this thing. We have to trust that there are really super-smart, awesome people out there doing the important work that needs to happen. But the reality is that we're going way too slow to make the important changes we need to make. And knowing that, how do you live with that reality and not lose your mind? And not break yourself into pieces?"

The Signs of Climate Anxiety

Eco-anxiety is not considered a diagnosable mental disorder for a variety of reasons, including the fact that it lacks clear diagnostic criteria. Yet there are common feelings associated with climate anxiety.

Guilt: For parents especially, it's common to feel guilty over the world we brought our children into or the world they are inheriting. So are feelings of guilt over our own carbon footprint or decisions we have to make that are better for us but bad for the environment. (A friend recently expressed regret over dental floss. She teaches her daughters that flossing is important for the health of their teeth, yet floss is a wasteful onetime-use product that ends up in a landfill. In these moments, it's important to remember that no one person is responsible for the climate crisis.)

Anger: There are likely many targets of your anger, from politicians to the mom who always leaves her car running at school drop-off. Our kids are angry, too, and it's normal for them to occasionally be angry with us.

Powerlessness: The issue of global warming is huge. Addressing the climate crisis requires immense systemic changes to so many things—from the energy system to our food industries to the way clothes are made—that it is easy to want to give up. It's a problem that touches every aspect of our lives, and the solutions are almost always going to feel imperfect. What good is not using a plastic straw or even switching your electrical system to solar when many airlines fly planes without passengers to maintain their takeoff and landing rights? The airline Lufthansa, for instance, recently admitted to running 18,000 empty or nearly empty flights in a six-month period. When faced with that type of news, we can easily hit a wall and feel completely powerless over our ability to make any sort of meaningful change through focusing on individual actions.

Overwhelmed: It's also common to feel overwhelmed by all the small (and big) choices facing anyone who wants to make an impact: Should you go vegan? Should you get an electric car? Should you buy cauliflower if it comes wrapped in plastic? Should you call your congressperson today, tomorrow, next month? The to-do list can feel endless.

Obsessive: Constantly thinking about the climate is also a sign of climate anxiety.

Grief and sorrow: From seeing images of skinny polar bears on glaciers, to witnessing the sky turn apocalyptic shades of red and orange from wildfires, to hearing the news of people suffering from climate-related disasters, how can we not feel grief and sorrow over what is happening? (If you find that your grief and sorrow are getting too heavy and slipping into feelings of existential dread and depression, it may be time to seek professional help.)

Climate anxiety can also cause physical symptoms such as panic attacks, insomnia or restlessness, changes in appetite, even digestive problems and headaches.

In younger kids, signs of problems may present as increased fussiness, changes in sleep habits, and even separation anxiety. In older children, look for irritability, an increased number of conflicts, or not being interested in the relationships or activities they once were invested in. If your children are in school, it's vital for caregivers to check in with school support personnel to get feedback on how they are doing there, because some signs of trouble may present only at home or at school.

"Remember, children are constantly changing," says Dr. Carla Manly, a clinical psychologist, family specialist, and climate-anxiety expert. "The more we keep our [finger on the] pulse of how our kids are doing—are they overeating, undereating, on their phone too much, talking about feeling hopeless?—the more we can look for the signs we need to take action."

"It's very [normal] for kids to be scared of big weather," Dr. Abigail Gewirtz tells me. What parents should watch for is when those feelings begin to interfere with their kids' daily life. "For instance, if your kid hears about a hurricane coming and they don't want to go to school because they're scared that it's not safe. Or if your kid has nightmares about Grandma and Grandpa in Florida being flooded."

While signs of depression and physical symptoms of climate anxiety should be addressed with the help of health professionals, remember that anxiety isn't always a bad thing. Many of us (especially Americans) have this misconception that people need to be happy all the time and that if we're not happy, we're failing. But all emotions are important, even hard ones.

"Anxiety can be a positive thing, depending on how we use it," says Yakaboski. "We want people to feel their anxiety and channel it in a positive way. The anxiety is a messenger [telling us] that we need to talk about why we feel uncomfortable."

How to Deal with Climate Grief and Anxiety

So how do we help our kids process their own anxiety and grief while also attempting to process our own complicated emotions around the climate crisis? We start by making space for feelings. Identifying and understanding emotions are skills. And knowing your own emotions and helping your child know theirs takes practice.

"We garden, we collect our water—there are so many survivalist skills that I think will be important for our future," says Bigley. "But at the same time, I think a lot of the important heavy lifting that I need to do right now is emotional and [about] well-being. [Our kids] are going to see over and over again places that they love hurting. I think our number one job [as parents] is to support their emotional state, to create a safe space for them to feel their feelings and work through their feelings."

Getting into the habit of making space for your kids to process their feelings, and then inviting them to share their feelings after they've had a chance to think them through is an essential part of building resilience and critical thinking skills. It's a necessary brief pause.

And it's an important skill for parents and grandparents, aunts and uncles, and everyone dealing with climate anxiety and grief to practice as well. As parents, we don't want to shield our kids from our emotions about the climate crisis, but we do need to process feelings and emotions without them present.

"Put on your own mask first," says Gewirtz. "If you are particularly sensitive—you feel very strongly—and you have a partner, let the partner do the talking. Deal with your own worry so that you can put that aside when you're having the conversation and be there for your child."

Talk with a therapist, or your partner or friends, and work through how you're feeling. It's perfectly fine to say you're feeling anxious, but the challenge is to not dump that emotion on your child. That's why it's essential to work through your own feelings and keep working through them; dealing with climate grief is an ongoing process.

When we process our own emotions and stay calm, we're demonstrating to kids of all ages what resilience looks like.

"I'm really committed to living my talk while recognizing that our system doesn't allow us to live fully carbon neutral," says Yakaboski. "As a parent you're constantly making negotiations and value judgments, and I articulate that whenever possible. For instance, 'We're not getting AC in the house and here's why.'"

It's vital when dealing with climate anxiety, grief, and parenting in general to have open lines of communication with your kids. Use your conversations and feelings around the climate crisis as an opportunity to foster and strengthen relationships.

Take the time to ask them what they worry about and acknowledge their feelings while also sharing how you deal with your own feelings around the climate crisis. Don't just say they're going to be fine; say it might not be fine, but it will get better. And then make it clear that you'll work through it together and help them come up with and implement practical solutions.

We start by helping children manage their emotions, letting them know that all emotions are okay, and that you can feel multiple things at once. "Embracing the grief feels so critical, and normalizing the conversation and the emotions," says Yakaboski.

Here are a few ideas:

Ages 4 and Under: At this age your child likely has no concept of the climate crisis, much less anxiety over it—but this is a good age to help build the foundation for emotion regulation. You can reinforce their love of nature by spending time outdoors and helping them practice taking care of it. Watch how you react to your child's emotions. If you're quick to make them feel better, you may need to pause and remember it's okay for them to not be happy all the time. You can get them to talk about their feelings and introduce an easy meditation practice. Help them understand that they can feel multiple emotions at the same time. For example, they can be excited to go with Nana for lunch but also miss you.

Ages 5–8: Your kid is now likely aware of the climate crisis at some level, but they're still too young to really comprehend it. You want to continue to reinforce their love for and connection to nature and the importance of protecting it, but you also want to check in with them about how they are feeling. This is especially important if you think they're aware of any news about natural disasters. If you, like me, are a catastrophic worrier, it's even more essential that you be mindful not only of what you say to your kids but what they see and hear, especially if you routinely watch or listen to news in their presence. You don't need to hide what's happening, but you also don't need for them to be exposed to constant worries.

"My youngest takes things empathetically," says Colorado Moms Clean Air Force member and mom Stephanie Rice, whose children include a seven-year-old. "He's emotional with how he views climate change. He'll say, 'Mom, I'm scared about how it's going to be when I get older.'" Rice says when those moments come up she focuses on being honest but also truthful. "I'll say I know we're not being the best doers for the world right now but there are many people who are seeing that things need to shift and we're making that work happen."

It is important for children to see how you react to things. When they're this age, you should continue to practice asking about their feelings and self-care.

Ages 9–13: You may start to see some of the more recognizable signs of climate anxiety and grief in kids this age, especially anger. Keep the lines of communication open, stress that lots of people are working really hard to solve the issue, and share a few examples.

Start to talk to them about some of the things they love and how they can help protect them, whether it's a beach cleanup or a campaign around polar bears or reducing food waste at home or at their school.

"I'm the kind of parent who thinks you should talk to your kids about everything in a way that's age appropriate for them," says Briana Warner, the CEO of kelp company Atlantic Sea Farms. "It's not being gloom and doom. It's giving agency, it's giving power, it's taking responsibility, and it's being factual. I think that we have the chance to help our children be resilient and help them be the kind of adults who are able to think through change, think through systemic problems, and come up with solutions and be optimistic, but we have to model that behavior for them. I started this job when my youngest was eleven months old, because it was clear to me at that point that I wanted them to see their mom going to work every day for something a whole lot bigger and something really novel rather than just going to work."

Encourage them to find ways to help (rather than feel helpless). "[My daughter's] reaction is often 'Well, why don't we do something?'" says Alison-Konteh. "She'll say, 'Why don't we write the president a letter?' So we do."

Ages 14–17: You can let older kids independently find resolutions. Listen first, and then ask what they need help or support with. Keep validating their feelings. They're angry, and there's some truth in that feeling of *How could the generations before us leave this mess?*, but don't let them sink into doom. You can say, "I hear you, I'm sorry, you're not alone, let's figure out what we can together." You want to cultivate the experience of openness and coming to you.

"It might not be an overt 'Hey, let's talk about your feelings,'" says Bigley. "But it's constantly giving them chances to have different perspectives, to be around people who are doing good work. [Or] being in communities where other people are doing that work and knowing that they're safe. They need to know that there [are] people around who have their back."

Kids need to (and should) feel like their voices matter. We can support them if they want to make their voices heard and be involved in trying to change the world. And we can encourage them to take news breaks when it starts to feel too heavy.

"There's so much to worry about," says seventeen-year-old Clare Flaherty. "It feels like there is a problem every day, but I try and do the best I can, and I'm still very hopeful about the future of our planet."

Practice climate optimism when you can, by focusing on courage and hope. Look for positive climate news. I often stop myself from reading certain stories that I know will make me upset. I also try to fill my social media feed with people and companies that I know are making a difference and sharing solutions, like Warner's Atlantic Sea Farms, local farmers, and climate scientists. We can share these stories with our kids and help them find their own climate heroes.

It's a tricky line to walk. We need systemic change to address the climate crisis, and it is not our responsibility or our child's to solve it. But our kids (and we) need to know our voices can matter.

Part 2 Love

Whether your child could spend all day outside in any weather, or would prefer to be inside reading a book or scrolling their phone, or falls somewhere in between, there is a way for them to take advantage of the benefits of time spent in nature and around animals.

You don't need to live by a forest or even spend hours outside to reap the rewards of nature. To empower our kids on a planet that is growing increasingly hotter, we need to foster environmentalism, but that doesn't mean every child needs to grow up to be a scientist. I'll show you how to use nature, animals, and your child's curiosity to build connections with the natural world and a sense of responsibility for it that will last a lifetime, while increasing their resilience.

3 Nature

Twinkling lights flicker in our backyard, and I hear the patter of footsteps as my child, who is supposed to be sleeping, comes running downstairs excitedly, saying she can see fireflies just like Grandpa told her she would. *Can we go outside to look at them?*

Even if the request doesn't make me smile, I know it's one I cannot resist. Years of research, interviews, and personal experiences tell me that ten, even five minutes outside doing something as simple as watching fireflies fosters a connection to nature that will help protect not only the planet but my daughter.

Like many parents and grandparents and aunts and uncles, I've found that the arrival of my child, and the wonder she takes in the natural world, have played an immeasurable role in how I look at our planet.

"In so many ways [having] my daughter reconnected me to nature and to very foundational values about living," says Dr. Tamara Yakaboski of Colorado. "The wonder and the magic of the world and really seeing things through her eyes reawakened a desire to share nature with her."

If you asked me to name the most unexpected thing about parenting, I would say it's this: You get to be a kid again. Not all the time, of course. There are lessons to teach, simple ones like not picking your nose, and harder ones like what being a friend means. There are a million trips down the stairs, umpteen snacks to fetch, stacks of dishes, and loads of laundry. But there are also puddles.

As a child I'd spend a couple of weeks every summer with my grandmother in tiny Hoosick Falls in upstate New York. We spent days at the lake, had ice cream for dinner, and every time it rained, we'd take a walk. I'd forgotten about those rainy-day walks until my daughter started to experience the joy of jumping in puddles.

For little kids, puddles are fun—an excuse to jump and splash and, even with a good pair of wellies, get wet. They can be a lesson in the natural world: about how rain falls, where it collects, and what happens as the sun comes out and dries it up. These walks are a way for kids to practice movement: a game of "follow the leader," walking and running around puddles and jumping both in and over them. They're made for exploration, with a tossed rock or a stick to poke in to see the depth. And if we want to get particularly educational about puddles, they can be measured.

But for me, the magic lies in that instant our feet hit the water. The joy my child finds in it. The laugh it produces. That connection to nature. Families who feel connected to nature feel more responsible for protecting the environment.

"One of the reasons climate change worries me is that my children need outdoor time to play," says mom of two

Dr. Joellen Russell of Arizona. "The PlayStation is not enough. They need to be out. They need to see the sky. And the people who remind me the most often about that are my dogs. They tell me they need to go out and they want their kids with them. With our long summers getting longer and hotter, [we need] those reminders that we're in this together, that we need to take care of our physical selves, which means our actual sky, our time outdoors, our time to watch the trees change and leaf and grow."

An oceanographer and a climate scientist, Russell is a member of Science Moms, an organization that advocates for climate action and arms parents with the latest climate information. She spends her days working on supercomputers, robot floats, and satellites, and says her time outside really drives home the specialness of our planet.

"We've found thousands of planets orbiting other suns, but we only have this one that is blue, full of water, full of life," she says. "I'm really trying to look after this treasure that I'm going to get to enjoy for this lifetime so that my children will get to enjoy it too."

Some evenings my daughter draws my attention to the moon. We'll pause and stare at the beauty of it, and I find myself grateful and hopeful that the sense of wonder she finds in the moon will translate into a feeling that she's part of something greater than herself, setting the foundation for a life of finding awe in not only nature but environmental stewardship.

"It really touches my heart when my son notices the beautiful tree that we're walking by and how the sun is shining through it or the snowflakes are glistening on it," says Yatibey Evans. "To me, having somebody recognize that beauty is a key component."

Evans grew up in Alaska hiking, dog mushing, berry picking, bicycling, and really just being outside.

"When I was a kid, my dad used to make me be outside two hours a day. As a kid, that's hard, like, *What are you going to do?* But now I feel like it really created a great grounding for me and helped me to appreciate what being outside can do for my mental health, my physical health, my understanding of the environment."

She's tried to pass on that same grounding to her own children, even if it isn't with two hours a day outside. "I've tried to spend as much time outside as I can while also taking care of meals and making sure things inside the house are functioning, but also talking to them about the beauty of the land and what it can provide for us."

Time in nature is also essential to a child's physical and psychological development. Research shows that spending time outside in different seasons forces children to use their brains in different ways and supports their cognitive development.

"Connecting with nature can be a teaching mechanism," says Dr. Deja L. Jones, cofounder of Honeypot Montessori, a forest school in Newark, New Jersey. "Nature can be a co-teacher. Every day that a child is spending time outside and they're exploring—rocks, leaves, insects—they're learning something new about the environment they live in."

Nature schools, or forest schools—where children spend the majority of the day outside, no matter the weather—are becoming more popular in the US. (The original forest schools were started in Denmark in the 1950s.)

"[Most adults] walk past ants every day and don't pay attention to them, but for a child [an ant is] a new species or something to observe," Jones says. "Every time that a child is outside, they're learning something."

When Jones started her PhD program, she wasn't focused on environmentalism but on the school-to-prison pipeline and how it impacts Black children. However, she quickly became interested in how nature is interconnected to child development. "I started to look at how lead impacts child development, as I was teaching kindergarten, and that led me to how the environment of a community really affects children. There are garbage incinerators all around here. There are also a lot of factories and children being diagnosed with asthma. I got interested in the environmental justice movement in Newark and [how it could impact] the children I was teaching, and it made me shift [the entire focus of my] research."

That's when she decided to open the Montessori-style nature preschool. "Even though Newark has all these environmental justice issues, how can we still give children a healthy nature experience in the midst of all these things that are happening around them?" Jones asks.

Not every community has an accessible forest school, but many schools, like the one Yakaboski's daughter goes to in Colorado, where the curriculum includes a weekly two-hour nature walk, are trying to incorporate more outdoor time.

In the United Kingdom, the Natural Connections Demonstration Project studied the effects of helping 40,000 primary and secondary school pupils get out of their indoor classrooms and outdoors between 2012 and 2016. They found that 92 percent of teachers said students were more engaged with learning when outdoors, and 85 percent saw positive impacts on student behavior. That's in line with what similar studies in the US have found: that overall, students perform better when their studies include outdoor education. Yet many schools balk at conducting classes outdoors, although more and more parents want their children to have time outdoors.

"My biggest fear was that enrollment was going to be our biggest challenge," Jones recalls. But when Honeypot opened up applications for their first year, two-thirds of the spots were filled within three weeks, parents with children not yet old enough were already applying, and families were asking about expanding the program to elementary and middle school.

The Natural Connections Demonstration Project found that in the UK, 64 percent of schools take lessons outdoors more than once a week. More than eight out of ten teachers said they'd like to take their classes outside more often, but 45 percent blamed their inability to do so on the curriculum. In the US, a movement is underway to green public schoolyards and encourage living classrooms, where students and educators take advantage of the experiential learning that can occur when lessons are taken outside. Senator Martin Heinrich of New Mexico has twice introduced the Living Schoolyards Act, which, if passed, would set up a grant program to provide resources for schools nationwide to rethink their outdoor spaces. So far the bill has not made it out of committee.

As schools grapple with how to provide outdoor time, there are simple ways child caregivers can encourage more time outside, starting by modeling it themselves.

Think back over the last twenty-four hours or even the last week. How much time did you spend outdoors? According to the EPA, the average American adult spends about 90 percent of their time indoors. That equates to five hours or less outdoors per week. What about your children?

A study commissioned by the UK's National Trust found that children spend half the time playing outside— roughly four hours a week—than their parents did as children. And that's not just a British phenomenon. The average American child, according to multiple recent studies, spends about ten minutes of unstructured time outside per day.

In the first days, weeks, and months of parenthood, I felt a bit like I was being called into the principal's office each time I took my daughter to the pediatrician, even when it was just for a well visit. I was the good kid in school, and the pediatrician visits were full of anxiety for me. Am I doing this right? Is she thriving? Is she happy? Am I going to raise a good person?

From the beginning, my New York City pediatrician— who was a lovely blend of hippie aunt mixed with my favorite high school science teacher, sprinkled with a bit of stern New England grandma—encouraged me to take my child outside. "Every day for at least an hour," she advised.

Her prescription for outdoor time felt daunting. Yet as the first months of new parenthood turned into the first year, and then the coronavirus pandemic hit, our pediatrician's advice remained the same: "If ever there was a year to splurge on outdoor gear, this is it," she told me as I shared my dread over the winter of 2020–21. "Dress properly, and get outside." So we did.

My pediatrician understood that spending time outdoors isn't simply enjoyable; it's essential. I wanted to know why and how I could inspire a love of nature to benefit my child and the Earth.

"If you take ten minutes with your child outside, it can make a world of difference in the way they approach the day," Ashley S. Lingerfelt, a licensed professional counselor and nature play therapist in Georgia, tells me. "It regulates our heart rate, breathing pattern, and respiratory rate, and emotionally, it helps us think very clearly and openly."

Talking to Lingerfelt made some of my personal experiences click—why I do my best thinking outside; why in a bad season of life, long morning walks help center me for the day.

Lingerfelt suggests prioritizing nature time in the morning and evening, because it's doubly rewarding. "It helps to regulate children both emotionally and cognitively but is also a nice time for kids to connect with their parents or their caregivers, after a long day of activities."

My daughter and I start most days with time outside, no matter the temperature or the weather, even if it's just a walk around the yard or a sit in the swing. It's a nice way to slow down and begin our day. We also start to wind down most days with an evening walk or time in the yard before we start the dinner, bath, and bedtime routine. It gives us a chance to check on the apple trees we've planted or throw rocks into the river and talk about the day.

Of course, right now, outdoor time is easy for us. My child is young, and the lack of demands from school, after-school activities, and my own flexible schedule means we can pop outside when we want to, but I know that as she gets older, it will get harder.

"Our time outside has dwindled in some ways," says Yakaboski, whose daughter is now ten. "But she's outside in different ways." They prioritize outdoor time through summer camps, bike rides, and even organized sports.

Studies show that children who feel connected to nature arc often happier and more likely to share and develop strong friendships than children who do not, and children and adults who spend time outdoors are more joyful, more content, and less anxious.

It's something Alison-Konteh, who homeschools her children, has noticed. "It would be nice to say that every day they want to learn and they're ready to go, but sometimes they don't feel like it, and I'll say, 'Get dressed—we're going in nature.' Sometimes they'll fight me about it, but once they get out there and start running around, playing in the grass, and take off their shoes, they get grounded in nature and feel a lot better."

Creating that connection can be as simple as spending time with a tree.

They seem like such ordinary things, trees. Your eyes likely take in multiple trees on any given day without you even registering them. But simply put, trees, like the big old dogwood outside our house, are priceless.

"Wherever there are trees, we are healthier and happier," writes Qing Li, a Japanese medical doctor and researcher, in *Forest Bathing*.

Trees are also climate-change fighters. According to the US Forest Service, trees offset 10 to 20 percent of US greenhouse gas emissions each year. Don't think that we can plant our way out of the climate crisis, though. Trees store carbon dioxide only temporarily, and 2023 research by Barry Saxifrage in *Canada's National Observer* suggests that in Canada, forests have reached a tipping point and now release more carbon than they absorb. Yet in cities, trees also reduce the heat island effect. Cities often experience higher temperatures than outlying areas because city landscapes full of buildings and roads absorb and reemit heat from the sun, thereby causing the heat island effect. But trees provide shade and help cool pavement, roofs, and walls to combat the heat island effect. Additionally, trees help improve soil, water, and air quality. They create more permeable soils, essential for reducing erosion and preventing floods and also for helping to clean and filter water, a process that more than half of Americans rely on for clean drinking water.

Trees are good for us. They absorb airborne pollutants, which can help reduce respiratory illnesses such as asthma, throat irritation, and more. Studies show that spending even a short amount of time surrounded by trees can benefit our immune systems, and some research shows trees can also improve heart health.

Shinrin-yoku, which means "forest bath," started to become popular in Japan in the 1950s. It involves slowly walking through a forest while taking in your surroundings through all your senses. In 1982, Japan launched a national program to encourage forest bathing as a way to reduce stress and improve overall health. In 2004, the country launched the Japanese Society for Forest Medicine, of which Li is a founding member, to study the effects of forest therapy on subjects in different contexts and forests. One of the Society's studies found that walking in a forest lowered blood pressure, cortisol levels, pulse rates, and sympathetic nervous system activity, and increased parasympathetic nervous system activity (related to relaxation).

Today, forest therapy is a well-known practice throughout Japan, with more than forty accredited shinrin-yoku forests. There are organized tours to help guide people in forest therapy. Still, many participate in some form of forest bathing alone in the accredited forests or at national parks and other green spaces around the country.

Researchers in Japan and other countries have conducted numerous studies on the health benefits of spending time with trees. So far, every study has shown a reduction in anxiety, depression, stress, anger, and even sleeplessness. While research is ongoing, it's thought that trees can affect the very structure of our brains, making us better able to handle stressors.

Researchers in Japan and other countries have conducted numerous studies on the health benefits of spending time with trees. So far, every study has shown a reduction in anxiety, depression, stress, anger, and even sleeplessness. Though research is ongoing, it's thought that trees can affect the very structure of our brains, making us better able to handle stressors.

Spending time in nature doesn't have to mean going to some pristine natural park (unless it's easily accessible to you). It can be as simple as stepping outside to watch the sunrise or sunset, or walking through an urban green space.

Alison-Konteh and her kids, who live in New York City, are able to find nature within twenty minutes at different parks, and they go camping, although she says it took some time to feel comfortable doing that.

"When we first went camping, we had a scary experience," Alison-Konteh tells me. "There were Confederate

HOW TO FOREST BATHE WITH A CHILD

1. First things first: Step away from your phone, or at least turn it off. You need to be present.

2. Take a walk in a forest or nature perserve if you can, but a neighborhood with lots of trees will also suffice.

3. Walk slowly. Lose the stroller if your child is of walking age, and amble together through the forest or the neighborhood. The idea isn't to hike, jog, or exercise, but rather to immerse yourself in nature.

4. Pay attention to the world around you. What does the air smell like; what does it sound like outside; how does the ground feel under your feet?

5. If you meditate with your child, this is also an excellent time to practice.

flags everywhere at the campground and a lot of rhetoric about having a gun, and when you see those things in conjunction as a Black person, it can feel scary. There was no research that prepared me for what I was going to walk into. You can't really call and say, 'Hey, do people have Confederate flags up at your campground?'"

Instead of giving up, though, Alison-Konteh, who realized how much nature spaces can be segregated, connected with an organization called the Hood Hikers, which takes people (it's geared toward adults) to hiking trails they know are safe in Connecticut, New Jersey, and New York. Finding that community gave her more confidence to take her children hiking and camping.

"The biggest thing I've noticed when we go camping now is the girls being able to play outside," she says. "I don't have to constantly watch them, [which] can't happen in the city. It gives them so much agency over what they're doing and the chance to explore more."

Ways to Get Outside No Matter Your Child's Age

Go Sound Mapping

Think about your backyard, a park near your home, or even a bench on a street in your neighborhood. What comes to mind? Are all your thoughts visual, or do any of them include sounds? Today, we devote a lot of attention to visual observations, yet accessing other senses, such as hearing, can help us connect with nature in a different way. Pick a spot outside and sit in silence for a set period of time, close your eyes if you can, and take in the sounds you hear and where you hear them. Some sounds might be immediately identifiable, such as the breeze rustling a wind chime. Others will happen intermittently: perhaps the buzzing of a bee or birds chirping. And then there are sounds you might be able to identify only after staying and listening for a while. Sound mapping is an activity that helps all of us use our brains differently. It's especially important for children, and it can help with cognitive development. It's also a way for them to connect with the world around them and teach them about seasons and the senses.

Additionally, for all of us, sound mapping provides a respite from our fast-paced world—a chance to unplug and connect with nature. It might feel strange the first time you participate in sound mapping, but over time it can become a form of meditation, a way to feel more centered in the world. With kids, you can also tie sound mapping into learning about the seasons by doing it once a week or once a month, slowly collecting observations that you can compare and contrast.

Talk About It

This is an especially important tactic with tweens and teens, because explaining the physical and mental health benefits and how they can foster a love for and desire to protect nature can help them prioritize time spent outside. No matter your child's age, though, talking about the benefits of being outside is a good way to help them want to spend their time outdoors, become curious about how it makes them feel, and practice emotional intelligence.

Grow Something

Gardening is one of those ideal kid activities because it can be as complicated or easy as they need. Little kids can help with planting, watering, and harvesting, and older kids can help decide what to grow and where to grow it, and even help build garden beds. You also don't need a lot of space to do it; you don't even need a yard. Here are a few tips to get you started.

Look inside your fridge. An edible garden should be full of things you actually eat. For us, that means lots of berries. At my daughter's request, we're letting the strawberry plants completely take over one of our garden beds this year.

Start small. A garden can be low maintenance, but it's never no maintenance, so start with a countertop herb garden (basil and cilantro grow easily) or a mushroom kit or a window box, and then if you have the room and interest, scale up each year.

Grow up instead of out. If you live in an urban environment or don't have a traditional backyard, every square inch of space is precious, so if you want an edible garden, grow vertically. Hydroponic or "smart" indoor gardens, like Gardyn, Lettuce Grow, or Click and Grow, among others, can also help you grow food in living rooms, kitchens, and even tiny studio apartments.

Enjoy. Not everything is going to grow successfully, and that's fine. You might end up with more lettuce than you'll eat, and that's also fine. For us, gardening gets us outside, actually gets my child to eat a few leafy greens that she'll pick in the garden, and shows her that we're part of the greater ecosystem, a good reminder to show respect to our habitat and treat it with the same care as we give our small garden.

Take a Child-Led Hike

Give your kid some agency over their outdoor time by letting them lead the hike. You can, of course, choose when the hike is happening and have veto power over where it's happening, but you should let them offer suggestions. After you've figured out those details, leave the rest up to them. They get to decide how far and how fast they go. You might not get very far, but that's not the point. Let older kids who might not care as much about being at the front of the line plan the whole activity. Again you might have to have some veto power or even have them pick from a list of suggested activities, but putting them in charge gives them agency over what is happening and lets them develop their own problem-solving skills, from building contingency plans for rain to what time you have to leave.

Schedule It

We're all busy, but outdoor time is essential for all the reasons we've talked about, so you may just need to schedule it. The more often you do it, the more of a routine it will become.

Create a Sit-Spot

A sit-spot is just what it sounds like: a place to sit. The idea comes from nature schools, where children are often allowed to pick a place outdoors that is special to them. Again, you can veto places that are unsafe, like too high in a tree or under a prickly rosebush, but the idea is to make the sit-spot somewhere that is easily accessible to your child's daily life. Encourage them to spend five to fifteen to even sixty minutes in their sit-spot. You'll need to start small, but as they spend more time in that spot, it can become a place for them to notice what's happening to the nature around them, to journal, or to simply think without the presence of screens.

Over the past few years, we've planted apple and dogwood trees in our backyard. I have no idea if any of the trees will live as long as the old dogwood that was at least 100 years old when we finally had to cut it down, but I hope that with some luck and help, they will. Looking at those baby trees that my daughter wants to check on at least three times a day, just to see if there are any more leaves, I know I'm helping to create a land steward. I look at her and the tree, and fifty years pass by. The trees are big, heavy with tart apples that my daughter's kids giddily scoop up as they run around the yard. The air is clean, the soil is healthy, and we've staved off the worst of the climate crisis.

◼⯐ Animals

The topcoat of alpaca fur is rougher than I expected. It's also dirty. Dried mud, bits of grass, and twigs rub against my fingers as I stroke a large black alpaca named Rosa. Yet there's no place I'd rather be in this moment than standing in the middle of a field interacting with these animals. The alpacas are as goofy as our family dog and just as hungry. When they realize my daughter is holding a basket full of food, they quickly follow and surround her.

We're decked out in wellies and raincoats. The summer mist is the type that will chill you to the bone, slowly and sneakily, and by the time you realize how soaked you are, it's too late to do anything about it. Yet we're giddy. All of us—my daughter (who isn't fazed at all by the enormous animals tracing her every move), myself, and a handful of strangers who showed up on a late July afternoon to feed the alpacas that call a gin distillery home.

"I think a lot of the time people think of themselves as separate from other animals and nature," says John Smalley, one of the directors behind Alpacaly Ever After, which rescues unwanted llamas and alpacas from all over the United Kingdom and gives them a home in the Lake

District. "Yet llamas and alpacas have very distinct personalities, and if you spend time observing them or being in close contact, you can't help but feel empathy and respect [for] them and, in turn, [for] nature."

Caring About Animals Means Caring About Nature

In Western households, pets are nearly as common as siblings, and while there's a lot we know about the health benefits of having a pet, there's much we don't know about child-pet relationships and what those relationships can mean for child development and environmental stewardship.

Animals have been my daughter's constant companions since before she was even born. Not alpacas (although how dreamy would that have been?) but rather dogs. My aging golden retriever, Vienna, sniffed her gently as soon as we arrived home from the hospital when my daughter was only two days old. That sweet dog got up with me for the middle-of-the-night feedings, observed diaper changes, and regularly plopped herself down near the crib or play gym throughout the day.

Vienna passed when my daughter was six months old, and it wasn't long before there was another dog, this one an energetic pup following my child's every move. Caring for both a puppy and a baby is no small feat, which is probably why many animal experts advise against it. The first two weeks were so awful that I suddenly understood how people can give up pets. Yet having my child grow up with a dog has always been a nonnegotiable for me, and we pushed through those early chaotic weeks.

Studies show that children who grow up with dogs tend to have sturdier immune systems—as someone who suffers from an autoimmune disorder, I wanted to give my child the best chance to avoid that. Yet the more I watched her with animals, the more I wondered what other benefits a dog could offer besides the ones we usually think pets teach children, like responsibility, kindness, and respect for boundaries.

Could having a dog help my child appreciate other animals, like alpacas, and even animals we tend to think of as creepy, like snakes and bugs? What about nature in general? When we talk about animals and nature, we don't often think of the animal companions we interact with daily. Still, I wondered, could having a dog help foster respect for animals and even help her process her climate anxiety—something I suspect she'll eventually have to do?

Many naturalists, including Sir David Attenborough, David Sobel, and Sy Montgomery, see benefits in cultivating children's relationships with animals from a young age. The idea is that something as simple and routine as walking a dog can help children develop empathy and promote stewardship of the natural world.

Even though I've watched my child take care of our dog, Caya, that felt like a stretch. Can taking a walk with our silly dog, Caya, help my daughter respect ants? Or the river?

"One of the things that my research has emphasized is that animals have the potential for teaching nurturing skills, but [these skills] are shaped by the adults and the parents and the people who are caring for a child," says Gail Melson. A developmental psychologist based in

Indiana, Melson has spent decades researching how nurturance puts down its initial roots in early childhood. She was partly inspired to study nurturance because she sees it as an essential tool for protecting the planet.

Nurturance, as Melson explains, is understanding the developmental needs of another, even when those needs might be very different from one's own or from one's expectations of what they should be. And it involves a commitment to doing what it takes to meet those needs. As we adapt to the changes the climate crisis will bring, we will have to live in a way that doesn't consider just our own needs but also those of our community and the world at large. How do we help children develop and practice nurturance? Pets who depend entirely on their human family for their physical and emotional well-being are many children's first introduction to nurturance.

At first, Melson's research with young children focused on caring for and nurturing the young of our species: babies. But what she discovered shocked her and raised some concerns. She found that by age five, children displayed substantial gender differences around caring and nurturing. "In the words of one of the kids we interviewed, 'Taking care of a baby is a mommy thing,'" Melson tells me. "I was surprised. We're in an egalitarian era, supposedly, but these kids had very different ideas. That concerned me, because being interested in nurturance throughout childhood is important for our species and ultimately for our planet. We can't afford to have half the population thinking it's a mommy job."

Melson and her research colleagues were curious about what, if anything, happens to that gender difference when looking at pets.

"We began to study the relationship between children and pets and found that first, there was a lot of nurturing and caregiving going on, and second, that the gender difference was gone. Boys and girls were equally involved, equally interested. And when we interviewed children, no child said, 'Oh, taking care of a dog is a mom thing.'"

For the first three years of my child's life, it seemed like every trip around the sun brought a new dog into my daughter's orbit: first Caya, the pup we acquired when she was nine months old, followed by my sister's pandemic dog a year later, and then my brother's puppy. Each one brought a new lesson in being gentle, picking up toys so the dogs wouldn't chew them, learning what they could and could not eat, and asserting herself around them.

Melson is quick to point out that parents and caregivers are the variable factor in the equation of pets and nurturing. A household where a dog is treated as a family member—for instance, where the dog sleeps on the bed and gets a birthday cupcake—will provide a very different lesson from one where the dog is a hunting or other type of working dog or, in extreme cases, an abused pet.

"Adults come in to help children understand what's appropriate," says Melson, adding that as a researcher, she makes no assumptions, and they don't yet have enough evidence to know whether nurturing a household pet can translate into nurturing wild animals or respect for nature as a whole.

Still, as more research is done, we continue to learn more about how a connection with animals can be important for respecting nature. A 2014 study found that animal experience, defined as having a pet or simply being around animals, led teenagers to be more likely to see themselves as important contributors to their communities. A 2016

HOW TO HELP YOUR CHILD PRACTICE NURTURANCE WITH AN ANIMAL

1. If you do have a household pet, have your child help take care of it. With supervision, even young children can be responsible for feeding a pet.

2. Talk with your child about how your pet has feelings and needs—for example, it needs a nap after playing.

3. If you don't have a pet, spending time with the pet of a friend or extended family member, or a school pet, can offer many of the same benefits. Many schools today have animals, such as chickens or a classroom rabbit, as part of the lesson plan, and parents can often volunteer to care for the pet on school vacations. You can also ask your child how they interact with the pet at school. Taking care of a pet lets students nurture their own creativity and practice their problem-solving skills. Melson's research has found that even robot toy pets offer many of the same benefits as a household pet when it comes to nurturance.

4. Find opportunities to have your child interact with animals, from local farms to experiences like feeding alpacas. Our local library has two resident kittens that my daughter loves to help take care of.

study in the UK found that children who had pets at home were more likely to think that animals have thoughts and feelings of their own.

How Pets Teach Children Empathy

"It's such a joy to raise dogs and kids together," Dr. Joellen Russell tells me. Russell is a climatologist and member of Science Moms. "The dogs help teach our kids responsibility, empathy, and about thinking ahead. And your dogs just remind you of what the most important things are, meaning time outdoors, time with your family, and time for joyful play. They've been lifesavers."

Russell and her family—her husband, two kids, ages fifteen and eleven, and two dogs—live in Arizona, where she is a professor at the University of Arizona. It might seem odd that a climatologist has dogs. Pets don't exactly lower your carbon footprint. A UCLA study found that what dogs and cats collectively eat in a year produces the same amount of carbon emissions as a year's worth of driving more than 13 million cars. Yet for Russell and many others, their benefits outweigh their carbon footprint.

"[Amid] the heavy lift on the science of discovery and then the policy and the arguing about regulation, et cetera, sometimes we forgot what we're fighting for," Russell says. "Dogs remind us. We're fighting to let our kids and our dogs play under the sun."

They live in a dog paradise from October through May, with cold nights and warm days when everybody lives outdoors in their backyards, and they walk all over the city with their dogs. But in the summertime, when the temperature often soars above 90°F, they have to limit the time the dogs spend outside, ensure the dogs stay hydrated at water stations set up throughout their house, and alter their daily routines for the dogs.

"We wouldn't be such an early-bird kind of family— we get up at five and walk them early—if it weren't for the fact that our dogs [need] that time, and it turned out to be a really good thing for the kids, too," Russell says. "They help with the doggy chores, including walking, groom-ing, and making sure the bowls are filled with water, and they get out and play early too."

Russell grew up with dogs, and while her husband didn't, having the dogs with the kids was always in their plans.

"I can't imagine a house without puppies," she says. "They keep us walking, they keep us outdoors, they're a good excuse for our neighbors to talk to us, because we're all out walking our dogs and we're accessible. Dogs make for a happier, more comfortable, family-oriented neighbor-hood, because you get to know everybody when you're out walking your dogs."

One of dogs' superpowers—and other animals' as well— is their ability to get us to look up from our smartphones and talk to people. Through our dog, my daughter and I have met and connected with people to whom we wouldn't normally give a second glance. These interactions are a chance for my daughter to practice her social skills, which I know will be essential to building stronger, climate-resilient communities.

"Your daughter is fearless." The woman on the side-walk pauses, not taking her eyes off my daughter, who is patting her gigantic dog on the back. "It's good to be fearless," she adds. Indeed, my daughter was and remains fearless around dogs and, to some extent, their humans. The animals seem to know she is a friend, and dogs we see on a walk will make a beeline to her. But her fearlessness toward animals is paired with care and caution. When we met this woman and her giant best friend, she said, "Hi. Can I pet your dog?" before carefully approaching.

She's learned—almost innately, it seems—to respect an animal's boundaries. And by seeking out experiences with other animals, whether feeding the goats at the farm up the street from us or feeding alpacas on a family vacation, she's continued to practice respecting boundaries, building community, and, I hope, developing nurturing skills.

Back at the alpaca farm, the alpacas appeared as wet as we were at first, but when I pushed aside the rough outer fur, the layer underneath was dry and cozy. My child held out the bucket, letting animals more than twice her size come to her to eat as instructed.

"[Seeking] to understand and respect other beings and how we all share the same environment can only have a positive impact on the decisions we make as the dominant species on behalf of the planet," John Smalley of Alpacaly Ever After tells me. "It improves empathy as we learn to understand the needs of another being, and impulse control by highlighting the need to be aware of our behavior, its effects, and how to adapt ourselves accordingly."

But children don't have to journey to the African savanna or even feed alpacas on a family vacation to connect nurturance skills learned from a family pet to the wider world. "I think the more relevant are the experiences that children have directly with the wild," Melson tells me. "Go out in their backyard; look at the worms."

Can Pets Teach Children to Respect Bugs?

Bugs can localize the natural world. So often when it comes to climate issues, we think of ice caps melting and polar bears instead of the environmental problems in our backyards. Bugs' role in ecosystems is essential. Many animals depend on them for food; they help to purify and aerate water and are vital to soil health. They're also in trouble. In early 2019, the journal *Biological Conservation* published findings that one-third of all insect species are in severe decline around the globe. If trends don't improve, we could face near mass extinction of all insects within the next century. Making that local connection is helpful for kids to see how they impact the environment right around them and how their animals do too. It's not always pretty. Sometimes you have to kill bugs.

"[Lanternflies] look so pretty, kind of like butter-flies," says Dr. Deja L. Jones, the cofounder of Honeypot Montessori, a forest school in Newark, New Jersey. "They're red with little bluish polka dots, but they're an invasive species, they kill crops, and they leave behind a black, moldy sap that is dangerous for us to eat. I think most people think they're just these cute little polka-dotted bugs, and I'm like, 'Oh, no, we have to kill that.'"

Jones, whom you remember from the "Nature" chapter, uses the outdoors as the classroom setting for the preschool. She says part of the focus of the school is on respecting and caring for animals. She's also open with the kids about explaining why it's important to kill those lanternflies.

"I think children are able to understand it better when they see how it's affecting other animals, people's homes, habitats," says Jones.

It's our job as parents to help our kids understand how the natural world is interconnected—the good and the bad. For instance, our dog, Caya, loves rabbits. She loves watching them, chasing them—but what would happen if she caught one? Well, in all likelihood, she'd kill it. It's vital that we connect the experience of our pets to those of animals in the wild. It helps children understand how everything is connected and how humans impact the environment and can ultimately protect or destroy it.

Before she turned four, my daughter had fed alpacas, goats, giraffes, dogs, donkeys, and cats; touched snakes; chased bunnies; and seen countless other animals. Interacting with any animal is a lesson in respect. We've all seen the headlines about dolphins that die as people take pictures of them or the buffalo that attacks when someone gets too close. At the beginning of our visit at the alpaca farm, the guide reminded us to let the alpacas decide if they wanted to eat, be stroked, or interact with us. How can that lesson be applied to humans? To nature at large?

A Moral Code

Pets teach empathy, which can help children develop their moral code, or what Melson calls a nature imperative—youthful concern about and commitment to conservation, habitat protection, and other ecological issues.

According to Melson and other developmental psychologists, children develop a moral code at a young age. For example, if your young child is playing with a cute red truck, and another kid comes up and grabs it, you child might react by saying, "That's not fair." That's moral code reasoning, and kids also apply it to animals from a young age.

Evidence shows us that no matter where they live, whether in a developing or highly industrialized country, in an urban or rural setting, children develop a moral code by which they value nature, care about animal welfare, and want to protect the environment.

"This is something that children don't have to have drilled into them," Melson says. "They are seeing and feeling, and they're concerned about what is happening in their environment as they get older. In a way, I think it's an encouraging finding, because it means it's not something that children are oblivious to, and it's not something that needs to be taught. But we do need to talk about the science underlying [environmental issues], the policies, and what can be done."

That was certainly true for thirteen-year-old Cash Daniels. Growing up in Tennessee, Cash has always been fascinated by animals. His family adopted a cat from the local animal shelter, but it's marine animals that have really captured Cash's heart. He has two turtle rescues and a seventy-five-gallon fish tank with beautiful freshwater koi angelfish. He goes to the aquarium near his home often, and when he's not there he can be found watching marine animals on the National Geographic or Discovery channels. While on a family vacation in Florida, he realized some of the animals he loved were in danger and knew he had to do something.

"I found a plastic straw on the beach, and earlier in the day I'd seen a turtle and manatee off the dock," Cash tells me. "I made the connection that the plastic straw could be harming them. So I knew I had to do something to protect those animals that I loved. When we got back to Tennessee, I got a couple buckets and I went out and started picking up trash."

Cash founded Cleanup Kids, and organized students worldwide to pick up a million pounds of litter before the end of 2022. "I want people to know that you're never too young or too old to make a difference," he says.

That moral code that kids are developing requires having empathy for others, something that dogs and other animals seem to teach kids naturally.

The summer my daughter was two, our second Covid summer, we spent a week in upstate New York at a cabin on the property of homesteaders who also run a nature school. When we arrived, the owners mentioned they had two dogs, and that although both were friendly, the one named Toby didn't like to be touched but would probably want to sniff us. I thought this would be a confusing concept for a two-year-old. *Let the dog sniff you, but don't touch him.* But my daughter surprised me—not questioning it, simply going with it. She continues to talk about it whenever we encounter shy dogs or even shy humans—"Oh, maybe he doesn't want to talk, he just wants to watch, like Toby likes to sniff but not be petted," she's said about a kid at the playground.

We Need the Emotional Support Animals Provide

When was the last time you talked to an animal? Watched your children stroke one? Anyone who has loved a pet knows that we turn to animals for companionship and comfort. While we and our kids know our pets don't understand every word we say, we still talk to them. Pets offer a nonjudgmental listening ear that can make us feel accepted and even less lonely. When Terry Barlow and Emma Smalley founded Alpacaly Ever After, they did so mainly for health reasons. Barlow was ex-military and thought that spending more time in nature with animals would be beneficial to his mental health. It was, and the organization has seen how visitors' interactions with the alpacas and llamas reduce stress, facilitate walking as therapy, reconnect humans with nature, and build self-esteem. But you don't need an alpaca or llama for that.

Study after study has shown that sitting quietly and petting a friendly dog (even if it's not yours) lowers blood pressure and cortisol levels—even more than spending time with a close friend. The next time you read a scary headline or your child has a question about wildfires or hurricanes, sitting next to and petting a dog might help you have those talks. They also might inspire action.

"The idea of having dogs and wanting them to be able to spend more time outdoors [as temperatures rise], when we know they can't sweat and that they have to drink water to keep themselves cool, is hard on dogs and hard on humans," says Russell. "They are telling us—just like our roofers and our agricultural workers and everybody who has to work outside—that it's getting hot in here. We need to take action."

5 Curiosity

"**W**hy does the yogurt taste better here?" We're eating breakfast at a hotel in Bowness-on-Windermere in England's Lake District, and my daughter's question catches our waiter off guard. "We're American," I say by way of explanation to him, although that is probably obvious, before answering my daughter's question with a question: "What makes it taste better to you?"

If you've ever found yourself googling how electricity works, why dogs lick you, or another question, thanks to your child, you probably already know that children are naturally curious. Exhaustingly so at times. On average, kids between fourteen months and five years old ask 300 questions per day. Scary, right?

If your child is older than five, chances are they ask fewer questions. By the time children are in early elementary school, the number of questions drops to an average of roughly seventy-two a day; by middle school, it could be down to zero. Why does that happen? And while it might mean you can finally finish a thought without interruption, is asking fewer questions good for your child?

The UN Development Programme has called curiosity a "secret weapon against climate change." When

I think of all the questions children ask, and how those questions can inform values and how they live, I can see why.

Maybe someday, my daughter's question about yogurt more than three thousand miles from home will lead her to make her own yogurt that lacks artificial sweeteners and preservatives, prioritizes flavor over yield, or at the very least sets the foundation for a willingness to try new foods and question how they are made and who and what benefits from them.

"I am a curious person, and I'm naturally curious about the natural world," says Dr. Erica Smithwick, climate scientist and member of the advocacy group Science Moms. "One of my passions is visiting beautiful places. We've had the opportunity to travel. We lived as a family in South Africa when I was on a Fulbright for seven months, and that was really transformative for my kids. I think that's when they finally started to say, 'Oh, Mom has a cool job,' but it's one of those things, too, where the apple falls very far from the tree. Sometimes when you're raising kids, they want to dissociate from the work you do. I actually don't care if they have a job in climate science. But I want them to be good humans on the Earth, and I want them to pay attention to some of the choices they're making."

To mitigate and adapt to the climate crisis, we will need deep thinkers. In other words, we need curious children who grow into interested adults. Adults who ask, "Why can't we ban fossil fuels?" or "Why can't we make airplanes fly on renewable energy?" Or even more simply, "Why can't I reduce my household food waste? And then find a way to do it." Curiosity, as Dr. Elizabeth Bonawitz,

Harvard University cognitive scientist and researcher, describes it, "acts as a kind of filter you put over the world to help the mind decide what information to attend to. It's a physiological response that helps drive action."

In other words, questions lead to change. It's not just that we need people to be curious and interested in science. Scientists like Dr. Holly Parker, director of the Schiller Coastal Studies Center at Bowdoin College, believe that one of the keys to effective climate-science communication and climate action is telling stories that lean in to curiosity and those childlike "why" questions.

"As climate scientists, we all feel the urgency of the problem," Parker wrote in a paper for the International Science Council, an organization that connects scientists with their peers and research organizations worldwide. "But the problem is not ours alone to solve. By embracing the curiosity of why, we can harness the power of inspired and invested policymakers, communities and individuals across generations and cultures to put our data to work. Together, we will tell the story of a more resilient, sustainable and equitable future."

Have you tried talking to other parents about climate change? Did you lean in to why you made a certain change, such as eating more plant-based foods, or your feelings about our rapidly warming world and what it means for our kids? Parker mentions a 2017 report in *Advances in Political Psychology* about how piquing someone's basic curiosity about science-based research may be the secret to depoliticizing hot-button topics such as climate change.

"I kept asking myself, *Why isn't the message getting through? Why aren't we doing the things we know we need to do? The science is pretty clear—just knock it off with fossil fuels—so why aren't we? Where's the urgency?* I really started thinking a lot about the role of storytelling and story listening—not just listening, but actually asking the questions that kind of pull the stories out of people," she tells me. "Make them comfortable to tell their stories so that we can understand why people hold the points of view that they hold and how that lived experience has brought them to a place where they might stand in opposition to something that would mitigate climate impacts in a community."

Curiosity helps us not only learn new things but also problem-solve and understand others. It helps us build relationships and create communities. Research links practicing curiosity with psychological, emotional, social, and physical health benefits. Multiple studies have shown that curiosity is associated with higher levels of positive emotions, lower levels of anxiety, and even higher academic achievement, improved technical skills, and better interpersonal skills in the workplace.

"I'm a big fan of curiosity," says Dr. Carla Manly, psychologist and climate-anxiety expert. "And I believe that we can have really lovely disagreements, because if you and I disagree on a topic and I say, 'Could you explain that to me?,' I might learn something."

Yet as we age, we tend to get less curious. Some of that is natural; after all, children ask all those questions to help them figure out how the world works and what things are. As they get older, they don't need to ask "What's this?" as much. But it's not just curiosity; several studies have found that people become less open to new experiences as they age. Why? One reason is that our current traditional systems of work and school aren't designed to encourage curiosity—and often they squash it.

School, in many ways, is about managing behavior and following rules. Sure, kids need to learn how to act in society. But too often education is less about learning and more about getting the correct answer. Research shows that when children begin school, they show far less curiosity than they did just a few months earlier while at home. And in the workplace, curiosity often takes a back seat to tending to day-to-day responsibilities, which can ultimately be a hindrance to productivity, not a benefit.

Fostering Curiosity in Our Kids and Within Ourseslves

According to Bonawitz and other cognitive scientists, you can't teach curiosity. Still, it is a trait that parents can and should nurture.

Being curious takes commitment; it takes perseverance and motivation. Being curious means you make a mess and get things wrong. Curiosity helps us learn new things, but it can be scary. You must be willing to do something unfamiliar, be open to new experiences and sensations, and often fail. Curiosity requires resilience, and resilience requires curiosity. But curiosity starts with questions.

It can be challenging, near impossible, even, amid all the required tasks of daily living, to stop and go on a deep dive into why yogurt in England tastes better than yogurt in the US, or to find out the favorite color of a California sea lion. However, studies show that kids are incredibly sensitive to how adults respond to questions. In other words, if you want kids to ask questions about the world, you need to encourage them and model your own curiosity—starting with promoting and practicing asking questions.

Instead of asking closed-ended questions like "Did you have a good day?," ask questions that require more than a shrug and don't necessarily have a right or wrong answer. "When we model healthy behaviors for our kids, they are picking it up [with their] nascent brains, they are growing," says Manly. "They are these little sponges that are becoming big sponges. Be really curious about your kids—curious about what's happening. Ask them about the lowlights of their day and the highlights of their day."

You know that age-old adage that there are no stupid questions. As adults, it can be hard to admit we have questions or don't know something. Yet asking questions is one way to practice curiosity, and it's a skill you only get better at with practice.

Admit to your children when you don't know something. Help them practice using their problem-solving muscles by sharing your own quandaries and asking them to share potential solutions. You can ask follow-up questions like "What do you think about that?" Or "Why do you think that?"

You can encourage their inquisitive nature with everyday activities no matter your child's age.

"You can use time spent outside to foster a child's natural curiosity," says Ashley Lingerfelt, counselor and nature play therapist. "Lean in to the world around you. Think about how things are made: How is this leaf made? How are these rocks made? That can be a catalyst for a ten-minute conversation [that piques] a child's interest to stay outdoors. Follow their lead to sustain their engagement."

You can also use the natural world to introduce kids to the cycle of life. "There are so many things outside—I mean, every single day—that are in the middle or the beginning or the end of a life cycle," says Lingerfelt. "Thinking about and viewing the world through that theme within studying life cycles, normalizing the idea of death and dying for children, but also using that to pique their interest in being outside and finding things to do . . . It might sound morbid, but there are plenty of ways to observe and be curious about dead animals, something as small as finding a dead bug and not really being afraid of those things. It's amazing, the conversations that can follow from just being curious."

The learning is deep when we connect our child's interests with a skill. For instance, if your child is intensely interested in collecting all the sticks they can find in nature, this is a prime time to provoke questions and observations. You can slip provocations in, like their characteristics, color, shape, size. Before you know it, you've covered one-to-one correspondence, color recognition, literacy, science, empathy, awareness, and more.

Spending time outside sets the stage for the motivation of curiosity in kids of all ages. Caregivers can grab hold of that and use the outdoors to help children develop cognitive, social, and emotional skills. A child who discovers and connects with slugs or other living things may develop a lifelong interest in conservation. Time spent outside is an opportunity to help children develop empathy and respect.

It helps to let your kids (and yourself) be bored. I let my child get bored. Sometimes, to get things done, I practice what writer Steph Auteri calls benign neglect.

USE A TREE TO INSPIRE A LOVE OF NATURE

1. Pick a tree in your yard, a favorite park, or on a walk.

2. Have your child name the tree.

3. If the tree is in your yard or at a park, set a blanket under it for reading or playing, make it the home base in kickball, or make it a meeting point. You want to find something that makes it a part of your child's everyday life.

4. Once a week, have your child touch it, listen to the sounds around it, smell it, and see if there have been any changes: Are there leaves? What color are they? Depending on your child's age, you could also use these weekly check-ins to start talking about the practical role of trees: Do any animals live in them? Does the air smell cleaner by the tree? What is the ground like near the tree? Can you see the roots? These questions can be jumping-off points to talk about the habitats trees provide or how they help clean the air or keep the soil from eroding.

Need some help figuring out how to talk about the tree? Use an app like LeafSnap or PlantNet Plant Identification to identify it.

We'll plan our day in the morning, and if it's a non-school day, it will likely include some independent (and ideally quiet) play. What my daughter finds to do often makes no sense to me—jumping over the cracks in our wooden floor, bringing her stuffies and putting them in random cabinets—but she is curious and exploring her world. You have to schedule boring sometimes, especially with tweens and teens who tend to have many activities. When kids get downtime, their creativity explodes.

You can use screens. Like many parents, I have a love-hate relationship with allowing my child on screens. I know that, at her age of four, the American Academy of Pediatrics recommends limiting her screen time to one to two hours daily. I also know that some days, allowing her to watch TV is the only way I can take a work phone call. And not all TV is bad; some children's TV programming can help teach kids valuable life skills like problem-solving and social skills.

For instance, a study from Texas Tech University found that watching the PBS show *Daniel Tiger's Neighborhood* resulted in greater emotional recognition, empathy, and self-efficacy in preschoolers, while television programs like Sir David Attenborough's BBC series *Wild Isles* have inspired conservation plans. Children's programming can even help teach children about the climate crisis and honor their emotions around it. Shows like *The Octonauts* (where undersea explorers help rescue sea creatures) and *Wild Kratts* (where two brothers meet various wild animals) do that, as do Disney movies like *Moana* and *Frozen*, which indirectly speak to our connection to climate change. Caregivers can use those shows to enter the conversation about the climate and continue to spark curiosity about the natural world by asking kids questions about what they observed and how they feel and to talk about actions.

You can cultivate older kids' curiosity by suggesting they find documentaries or podcasts that address the questions they have around the environment. A teen interested in vegetarianism can find plenty of documentaries about the agricultural system. Parents can follow up by asking what they observed and if they think the source is trustworthy, and then suggesting ways to dive deeper into research. For instance, a child who loves to watch documentaries about the ocean and marine life can explore the average water temperatures sea turtles need and follow their migratory patterns.

Read. Take your children to the library and ask what they want to learn. It's been sharks recently in our house, and we've gone through many shark books. Find books on topics you're curious about as well. It doesn't have to be nonfiction—diving into a novel set in a different place or time or with a character who is different from you or your child can also cultivate curiosity.

Accept messes and failure. Every part of life doesn't need to be perfect. Part of letting my kid be bored means that I will sometimes come upon a mess of paper cut into tiny pieces or paint splattered on the wood floor. I take a deep breath and let it go; we can clean it up. I remind myself that she was exploring and engaging her creativity, and that's what matters. Let kids help pack suitcases and see how many clothes they can fit in. So what they unroll the toilet paper roll: Let them try to roll it back up. When it comes to schoolwork, try to take a step back and not make it all about grades (some of that, I know, is a privilege), but curiosity requires failure.

Change your routines. I love routines. I'm a stickler for them, but small changes in patterns force our brains to think differently, which creates curiosity. With young kids, it can be as simple as changing the cup they usually drink out of; with older kids, it could be taking a different route home.

Use shopping to spark curiosity. We spend a lot of time at the store talking about the things on the shelves, what they are made of, and how they are made. As my daughter gets older, we'll talk more about whether something is good for the planet, good for us, and good for the people who made it. It's an opportunity to flex her curiosity in a new way and start making her aware of her privilege and the responsibility that comes with it.

Travel

I can still remember the feeling of going to the post office with my mom at twelve or thirteen to get my first passport. It was scary and thrilling; the whole world felt open to me. Travel is often cited as one of the go-to ways people fuel their curiosity, and many parents of young children dream of showing them their favorite places. I know I do.

"I have the longest bucket list [and] am deciding if I'm going to keep [the items] or let some of them go," travel journalist and parent Michele Bigley tells me. "If we want our grandkids to be able to enjoy the world, flying [into a locale] and renting a big car and staying at a big resort, or [renting] an Airbnb that actually makes it really hard for local people to afford to live there—we have to measure all of those costs: the larger costs to these places we love and what our larger footprint is."

Aviation currently contributes around 2 percent of overall global emissions, increasing more than those from any other form of transport. A flight from the West Coast to the East Coast of the US produces at least one metric ton of carbon dioxide. That's on top of the average amount of carbon monoxide produced by each human—about 5 metric tons a year. (Though, if you're American, Australian, or Canadian, multiply that number by three. It's also worth noting that in the US, just 12 percent of people take 66 percent of flights. In France, 50 percent of flights are taken by 2 percent of the population. In the UK, 15 percent of the population takes 70 percent of all flights.)

Yet there are so many places I want to see, so many more I want to show my daughter, and so many things I think you can learn only through travel. Can I show my daughter the world without doing unnecessary harm? I don't have the answers. I don't think anyone does. But I admire how Bigley considers the benefits and the consequences, and I think she does a good job of finding a middle ground.

"I love how travel connects my kids with other humans that are very, very different from them," Bigley says. "It shows them that our way is not always the best way and makes them more adaptable. It also makes them realize they can do a lot of things they didn't think they could do, and it makes them more resilient, because things go wrong all the time when you travel."

Bigley is a fan of regenerative travel. If you think of sustainable travel as leaving a place the way it was when you arrived—which, let's be honest, isn't always great—regenerative travel is not only attempting to make it better but, ideally, letting the people who are there, who have some sort of indigenous ties to that place, tell you how to make it better.

"It's the difference between packing up your trash when you go to the beach and cleaning up all the trash that you see on the beach," Bigley says.

So how can we travel?

Go off the beaten path. Can you travel during the off-season or to a less-popular destination? Doing so can help lessen the impact of travel on more heavily trafficked destinations. This can be hard, especially when your schedule is dictated by the school calendar, but if you can, consider the time of year and where you're traveling.

Choose local, plant-based meals. You can significantly cut your travel footprint by choosing local ingredients and/or plant-based meals.

Opt for public transportation. Traveling to your destination by train or bus can make a difference. And remember to think about how you'll get around once you arrive at your destination. Can you take public transportation instead of hiring a car or taking a big tour bus? Or can you rent or borrow a car from a local family for a day or two when you need one?

Be resource-conscious. As you do at home, switch the lights off, keep showers short, and avoid plastics by packing a tote bag and a reusable water bottle.

Stay longer. When my daughter was two, thanks in large part to my ability to work for myself and remotely, I started taking her on a monthlong summer trip. I picked a destination; one year, we would fly, and the next, we would drive. Now, obviously this isn't feasible for everyone, but staying in one destination for an extended period, known as slow travel, with longer trips instead of several smaller ones, has multiple benefits. It reduces the carbon emissions associated with getting to and from destinations; during the stay it reduces turnover emissions at a hotel or rental; and extended stays may also benefit local economies if you are mindful about spending your travel budget at local businesses.

Be mindful of your surroundings. Be conscious of your impact on local communities, such as where you stay. For instance, many short-term rentals are pricing people out of homes. At the same time, major hotel chains often send more money out of communities in a process known as tourism leakage. Look for locally owned guesthouses or independent, locally owned hotels, and extend that mindfulness to your day-to-day activities.

"Think about what you are bringing to the table," Bigley says. "We can't just expect a place to welcome us with open arms. We need to realize that places are hurting, so what can we do to support these places the best way we can?"

Fly less. There's no way around it. To travel sustainably, we need to fly less frequently. Yes, work is being done on adopting sustainable aviation fuels and new technologies, including electric planes. Still, in the meantime, we need to reduce our flights. Research published in a 2019 report titled *The Future of Urban Consumption in a 1.5°C World* found that one way individuals can reduce greenhouse gas emissions is to keep flying to a minimum (no more than one short flight, defined as less than 1,500 kilometers—about 932 miles—every three years and one long-haul flight every eight years). Put another way, a flight from Providence, Rhode Island, to San Diego is more than 4,000 kilometers (2,485 m). Activists like Thunberg have given up air travel entirely, and others have signed no-flight pledges or are limiting the number of flights they take. Don't be fooled by "carbon offsetting." It's a Band-Aid over a much bigger problem, designed more to make us feel better than to actually do good.

Our curiosity and our children's curiosity are key to building individual, community, and societal resilience. The more curious we are about our neighbors, the trees outside our door, and where and how our clothes are made, the more willing we are to understand and empathize with others. Curiosity can help us build community and, as Parker suggested, helps to depoliticize topics of disagreement, because it can be helpful to think about how others' experiences may shape their opinions.

Curious kids ask questions, try new things, and grow into engaged, inquisitive adults. Curiosity, no matter your age, increases a brain's plasticity, giving it more power to adapt to and handle stress and large-scale problems such as the climate crisis. Our children need these abilities to mitigate the worst of the climate crisis and create a more just, sustainable world.

Part 3 **Build**

Humans are driving planetary warming. That means humans can change it. In these next chapters, you'll find ideas and implementation strategies for the actions individuals can take to make a difference.

It starts with problem-solving. It's crucial that the climate-crisis generations have the skills they need to be effective problem-solvers. As parents, we can help our kids work those muscles.

Then we'll look at key ways individuals can reduce their greenhouse gas emissions, starting with the food they eat. We can't avoid the worst effects of climate change without cutting food system emissions. But what can individuals and families do? Do children need to follow a particular diet to enact change? Do parents need to lead by example? Is ice cream still okay? Chicken fingers? Salmon, a fish my daughter loves, but could be extinct or unavailable before she's my age? You'll learn how to determine the answers to those questions and more for your family.

Finally, we'll problem-solve around consumerism, and look at how individuals can help create the cultural shift that will need to happen for us to mitigate and adapt well to the climate crisis.

Nothing will change without systemic change. But individual action matters too—if only to keep us hopeful.

6
Problem-Solving

When my daughter was a toddler, her voice would ricochet off the walls of our mudroom: "Mama can't do it; Mama can't do it," she'd repeat over and over. She'd insist on zipping up her yellow raincoat herself, even though she hadn't mastered the mechanics of getting the slider onto the teeth. I'd take a deep breath, attempt to use all the patience, calmness, empathy, and stability the gentle parenting moment required, and ask if we could do it together. Little did I know then that "Mama can't do it" would become a frequent refrain for her and that it might be a good thing.

Based on some fundamentals of the RIE (Resources for Infant Educarers) style of parenting popularized by Janet Lansbury, the term *gentle parenting* was coined by parenting expert Sarah Ockwell-Smith in 2015. Also known as respectful, intentional, or mindful parenting, gentle parenting is a philosophy that centers on understanding, empathy, and respect, and builds a supportive partnership between parent and child. It's a willingness to meet children where they are developmentally and consider that children are individuals with their own likes and needs.

Gentle parenting isn't easy. Staying calm and collected while simultaneously validating emotions, especially in high-stress situations, is very challenging. It also can be hard to get buy-in from co-parents or other caregivers. I was at one of those class birthday parties with my daughter—you know, the ones where the whole class gets invited? A few of us were trying to also make it to another class birthday party afterward. One child had a meltdown about leaving, and the response from some of the older adults in the room was "How's that gentle parenting thing going for you now?" We laughed about it when we got to the next party, reminding ourselves that despite the challenges of gentle parenting, you're in it for the long haul, to help your child grow into a resilient and emotionally healthy adult.

I don't think it's possible to stick to any one parenting philosophy but one important element of gentle parenting that I like most is that it encourages parents to mess up in front of the kids. Our missteps can be an opportunity to show kids that it's okay to make a mistake, and model how they can recover from a mistake.

This is also an essential part of problem-solving.

When you're in the moment, it's tempting to remove whatever obstacle is facing your child, whether it's a rain boot on the wrong foot, soccer cleats forgotten at home the day of the big game, or even forgetting to sign up for college courses before the deadline. It's often easier to solve their problems and move on, especially in a world where we're all so busy. Yet every day is a chance for children to flex their problem-solving muscles, a skill that will be critical for children and the adults they'll become while living through the climate crisis.

At this point, we can't really solve climate change. We can mitigate it and adapt to it. The reports from the IPCC, for instance, list lots of potential big-picture ways to do that through policies like renewable energy and adaptation strategies like building seawalls. We know we must reduce emissions. And to do this, we need to solve a whole host of other problems, like getting corporations to reduce their energy consumption, governments to commit to clean energy and end fossil fuel subsidies, and our neighbors to stop using gas-powered leaf blowers.

Even as we work to avert the worst potential impacts of climate change, we must become more resilient to its now unavoidable effects. As the world grows warmer—and make no mistake, it will—mitigating the climate crisis isn't just about cutting carbon emissions. It's also about protecting people from harm and figuring out how to live joyfully and thrive while we adapt to our new reality. To do it, we need climate resilience and good problem-solvers.

People with strong problem-solving skills are resilient. Think of all the things that go into problem-solving. There's identifying the problem, breaking it down into manageable parts, brainstorming solutions, evaluating the feasibility of the solutions, trying a solution, monitoring the results, and adapting it as necessary. You need to ask questions, be curious, gather information, and be able to build community. You need to be flexible and adaptable, and you have to be able to fail and try again. In other words, all the skills that make it possible to solve problems, including analyzing the situation, creating possible outcomes, and applying solutions, are the same skills that help people become resilient. It's no wonder research shows that creative problem-solvers are happier and may even have stronger relationships.

No matter their age, it's not always easy to let your child problem-solve. It means occasionally being late when you're waiting for a two-year-old to zip up their jacket, or letting the teenager who stays up too late on a school night risk sleeping through a test. Allowing kids to problem-solve means you have to let your children take risks, fail, and then deal with the consequences and the disappointments. Or the muddy clothes.

These days, our children grow up in very controlled environments. Some of that is, of course, necessary. Still, over the years, we've managed to take risks out of a lot of childhood, and with this, opportunities for children to practice their problem-solving skills. Play is structured, with planned playdates where parents hover, trying to make awkward small talk; sports are organized and super-vised; even playgrounds have been reduced to risk-free environments with cushy artificial grounds, surrounded by fences, the seesaws and merry-go-rounds relics of the past.

"The playgrounds here are better," my then three-year-old said matter-of-factly. We were in England for the summer, and while I agreed, I was curious what she meant. "They're more fun; we don't have these things at home," she replied when pressed, pointing to the zip line she had just been on and the steel merry-go-round. Over the past decade, the United Kingdom has made an effort to make its playgrounds riskier. During our summer we found playgrounds with merry-go-rounds, seesaws, zip lines, and lots of grass surfaces and trees to climb.

These playgrounds reminded me of the adventure-style playgrounds I had reported on previously, which were popular in the 1960s and 1970s in the US. Those playgrounds, a few of which still exist around the country, used custom-designed mounds, pyramids, mazes, and water features to encourage exploration and creative play. Over the past decades, a number of places, including the United Kingdom, Sweden, Canada, and Australia, have been rethinking playgrounds such that they provide some risks, which early childhood experts say is part of helping children grow into resilient adults.

"A lot of times in my work, I see that parents are afraid to allow their children to take risks, because parents want to protect their children from disappointment or failure or pain," says Ashley Lingerfelt, professional counselor and nature play therapist. "I understand that, and obviously we don't want our children to do anything unsafe, but there's a big difference between participating in something that's unsafe versus something that has risks."

For young children, being outdoors and playing in an unstructured environment is full of opportunities for risk-taking and problem-solving. Unstructured play is open-ended with no goal. It's time for children to rehearse and make sense of the world. Balancing along the edge of a creek bed and climbing a tree help with gross- and fine-motor development and sensory stimulation—and they also carry a bit of risk.

"How can a child know why their parents say, 'Don't touch that rosebush' if they have never experienced the prick of the briar?" Lingerfelt says. "It is not harming them, but it's something, in my opinion, [that is] very important for them to have the opportunity to learn."

You might not want your child to roll around in a rosebush. From a preventive standpoint, Lingerfelt recommends allowing our children to play in a space that is inherently safe.

"Would I say, 'Hey, let's go along the edge of this really steep fifteen-foot riverbank and just explore'? Maybe not. But whenever we think of risk-taking, generally speaking, is a shallow creek safe? Yes. The worst that could happen is maybe they fall over or they trip, but their life is not in danger."

If your child does fall into a creek, or, in my child's case, the pond at nature school, it's important to see how they react to getting wet. To watch how they handle adversity.

"Oftentimes the child recovers pretty easily without our help," Lingerfelt says. "Now, that doesn't mean we ignore our child, but we can say, 'Oh man, you got wet,' versus saying, 'Oh, you fell.' See how that's different? We don't need to automatically assign meaning to the child's experience. We need to let the child experience the meaning for themselves, or create the meaning themselves."

When we had to have trees cut down in our yard, I created a jumping path with the stumps. They'll eventually decay, but for now, it's a playscape. When I was placing them, my dad kept moving them closer together, thinking they were too far apart for my daughter. Still, I kept pointing out that she'll be fine if she falls or she'll find something to hang on to to get herself across. She's developing the tools to conceptualize a solution to a problem. Whether it's rain on a planned afternoon to the beach, a lost sports game, or a canceled flight, little and big disappointments can help kids practice problem-solving skills and build resilience—but only if children experience them.

Dr. Robert Brooks, coauthor of *Raising Resilient Children*, says that shielding kids from disappointment prevents them from learning essential skills, such as the fact that they have the tools to get over disappointing situations. That isn't to say that you shouldn't help them zip up a coat or solve a problem with a friend but that you also have to help them learn to ask for support and communicate what they need.

Ages 4 and Under: Play and Emotion

Play is one of the best ways to start giving young kids the skills to analyze problems, imagine possible outcomes, and apply solutions. Even if you don't have an adventure playground nearby, time spent outside can help foster that love of nature as well as problem-solving skills. As Lingerfelt mentioned, let natural consequences do the teaching. To solve problems and find their place in the world, kids need to be able to take risks and practice their creative skills.

Encourage your child to communicate their emotions and needs. "Coach them in a really positive way, with basics like using *I* messages, which most adults have trouble with," says psychologist Dr. Carla Manly. "Use *I* messages—'I feel this, I feel sad, I feel hurt, and this is what I need.'"

Continued on page 102

HOW TO BUILD PROBLEM-SOLVING INTO CHILD'S DAILY LIFE

Gardening is full of opportunities for problem-solving, from figuring out what to grow and where to grow things to troubleshooting pests and watering issues.

Organized sports: Team sports, even with hovering adults, require quick decision-making and help teach children how to solve problems and adapt.

Cooking, baking, and meal-planning: You've got to be brave to cook with kids, especially young kids. And maybe a little foolish. Nothing will go as planned. It'll be incredibly messy. You'll likely forget an ingredient or a step. Yet the result will be pretty sweet. It's basically a metaphor for life. But with kids of all ages, looking in the fridge and pantry and coming up with a meal encourages problem-solving. With older kids, cooking and baking are full of tiny problems to solve in the moment.

Grocery shopping: A trip to the grocery store is a great time to talk about values and hone problem-solving skills. My daughter is at the age now where she understands that plastic isn't good for the environment, but she also understands that sometimes there is no other option. I'll ask her to scan the shelves for applesauce in a glass jar. When she sees that there isn't any available, we problem-solve: Should we buy a bigger container so it's less plastic overall? Is there something else we can do to make up for the plastic we're buying?

Home DIY projects: We live in an old house. There are many things to fix, whether it's the wiggly doorknobs, a shingle that falls off, or a rotted board. The home DIY projects require persistence, strategizing, resilience, and often lots of trips to the store.

Traveling, camping, and hiking: Traveling isn't just a tool for inspiring curiosity, it's also a chance for practicing problem-solving, just as camping and hiking are. Lots of unexpected problems come up that require adaptation. Let your kids help find solutions to common issues like flight delays, lost luggage, where to set up a tent, and more.

Managing money: You may have mixed feelings about allowances; I know I do, but allowances can help teach kids about delayed gratification, budgeting, and how to plan and save for things they want. All are part of problem-solving.

Continued from page 99

Work on helping kids identify their emotions. A crucial part of problem-solving and building communities is the communication part of emotional intelligence, saying, "I feel joy, sadness, anger, fear, disgust," and not assigning positive or negative judgments to those feelings.

Ages 5–8: Practice Finding Solutions

As kids get older, problem-solving can evolve into finding solutions. It can be as simple as grocery shopping with you. If they really want juice, for example, can you ask if they can find it in a bottle that can be recycled? Or if they see something they want online or on TV, say, "Do you think you might have a friend who has that and wants to share?" or "Let's look into secondhand stores." Give them other ways to get what they want while also seeding the idea that we need to consume less.

Ages 9–13: Independence and Accountability

Continue to give them more independence and ask things like "I don't know, what do you think?" or "Are you thinking or are you stuck?" to give them the chance to decide if they need help instead of simply rushing in to solve the problem.

"I've been thinking about this a lot for the last six months," says Dr. Tamara Yakaboski, the educator and climate-crisis practitioner based in Colorado, about her ten-year-old, "realizing that I want her to be more independent and give her more accountability and freedom but also recognizing that because of Covid, we haven't given these kids a lot of skills." She mentions sleepaway camps and sleepovers, which her kid had for the first time in fourth grade.

"All of the fourth graders slept over, and she actually hasn't had any more. I don't know about you, but I grew up with sleepovers."

I did, too, and if you did as well, you probably had to figure many things out at the sleepover, from who was putting their sleeping bag next to whom to negotiating about which movie to watch or whatever activity was happening. In resolving conflicts or understanding the needs and preferences of others during sleepovers, children can develop empathy and the ability to see situations from different perspectives. These skills are essential for effective problem-solving, because they enable them to consider the feelings and viewpoints of others.

Ages 14–17: Model Problem-Solving

Is it your problem or your child's problem? Continue to let natural consequences do the teaching; for instance, if they don't study, they'll get a lousy grade. But then work with them to come up with a solution. Break down problems together. For example, if they always forget their soccer cleats at home, brainstorm ways to avoid this, and help them weigh the pros and cons of possible solutions.

Talk about and model your own problem-solving, such as deciding what's for dinner, being nervous about a presentation you're giving at work, or, when you lose your patience with someone, being vocal about what you'll do differently in the future. This shows our kids that when you make a mistake, you're going to recover from it. It's beneficial for children to watch their parents make mistakes, admit them, and right their wrongs. No one is perfect, but if you solve your problems calmly and rationally, your child will learn to do the same.

With kids of all ages, you can use their concerns about the environmental issues facing their generation to boost problem-solving skills. Involve them in activities like conserving energy and figuring out how to reduce waste.

"On an individual level with kids, it's nice for them to be able to feel like 'Oh, I can actually do something.' Being able to do something that you can physically see is very powerful compared to advocating, where you're not necessarily going to see the result right away. Both are important, but I think each can feed the other," says Brooke Petry, the South Philadelphia mom and organizer with Moms Clean Air Force.

Creativity, curiosity, adaptability, and flexibility are critical to developing problem-solving skills. Those traits are also integral to handling climate dangers and changes, and parents and caregivers can help their children hone these skills in small moments.

■■■7■ Food

Food has superpowers. It's sustenance, yes, but our souls are also wrapped up in it. It's the ice cream party with friends before the start of a new school year, the ginger tea a mom makes every time her child is sick, the corn bread recipe passed down through generations. Food is also how most of us can make the most significant impact on the climate crisis. No matter our child's age, food offers us a way to talk about the environment, foster respect for nature and animals, cultivate curiosity, and practice problem-solving.

"Cooking with my kids, teaching them how important it is to connect to our food, and talking about issues around who has access to food, who has access to land, and why that is and why that should change makes me sad, but also really hopeful that we'll have young people who are coming in and caring about a lot of these issues," says Alexina Cather, mom of three and the director of policy and special projects at Wellness in the Schools.

It's no secret that food production significantly contributes to climate change. One of the most recent and comprehensive assessments of the food system's overall role in the climate crisis suggests that in 2018 the global food system was responsible for 16 billion metric tons of greenhouse gas emissions, or a third of all global emissions. From agriculture practices, to the packaging and transporting of food, to agriculture-driven deforestation, the analysis, published in *Environmental Research Letters*, reinforces what many climate scientists and food system experts have been saying for years: We can't avoid the worst effects of climate change without cutting food system emissions.

Unless you own a farm or a food business, you're likely wondering what that means for you. A lot. According to numerous studies, food accounts for anywhere from 10 to 30 percent of a US household's total carbon footprint.

How to Reduce Your Family's Foodprint

Reducing food waste, teaching our kids about where food comes from, saying no to plastic, and eating more plant-based foods are all things we can do to reduce our food carbon footprint, or foodprint, while raising climate-resilient eaters. Here's where to start.

Reduce Your Meat and Dairy Consumption

The single most effective dietary change you can make to reduce the carbon footprint of your food is to reduce your meat and dairy consumption. In 2014, a study found that vegetarians had about half the food-related carbon footprint of meat-eaters, and a 2023 study of more than 55,000 people and 38,000 farms in 119 countries found that

YES, YOU CAN EAT MEAT

Conventionally raised beef is the least environmentally friendly meat, in terms of both the land it requires and the greenhouse gases it produces. But there are livestock farmers raising animals sustainably, managing their land to reduce the effects of greenhouse gases. Use these tips to guide your eating choices.

1. Eat fewer animal products. Can you commit to going vegetarian one day per week? How about two or three? Can you be a weekday vegetarian or vegan?

2. Swap out beef for pork or chicken. Cutting your beef consumption by half and replacing that half with pork could save 20 percent in greenhouse gas emissions. Poultry has an even smaller climate footprint than pork.

3. When you shop for meat, look for grass-fed/pasture-raised animal products, preferably verified ones so you can trust the claim. Or, if you have access to one, buy your meat at your neighborhood farmers market or specialty butcher shop, where you can talk to the people producing it and ask them questions about how they treat their staff, the environment, and the animals.

SEAFOOD TOO

According to the UN Food and Agriculture Organization, seafood consumption has outpaced consumption of all other animal protein foods. And seafood consumption worldwide is expected to double by 2050. But like other animal protein foods, not all seafood is created the same.

Here are some tips to guide you when purchasing seafood.

1. Eat a wide variety of fish.

2. If you live near the coast (within 145 km/90 m), buy local, seasonal fish whenever possible. If you don't live near the coast, frozen fish is often a good bet, but you'll want to try to buy as close to local as possible. Look for the abbreviation IQF (individually quick frozen), meaning the fish was frozen within minutes and then stored to be shipped. FAS means it was frozen at sea, but it might have been thawed, reprocessed, and frozen again.

3. Buy American seafood, starting with wild-caught. The fisheries within the US are highly regulated.

4. Steer clear of imported seafood unless you're confident about where it's coming from. Countries have a wide range of standards around labor, environmental protection, antibiotics use, and more.

5. Buy from a fisherman, if you can. Find one using Local Catch Network's seafood locator. Otherwise, introduce yourself and your kids to the fishmonger at your supermarket. They can help you find a variety of fish to buy and answer your questions.

plant-based diets produce 75 percent less heat-trapping gas, generate 75 percent less water pollution, and use 75 percent less land than meat-rich diets—those that include at least 100 grams of meat daily, the equivalent of one steak about the size of a deck of cards.

"My kids really like meat, but I talk to them about the impact of meat production on climate change," says Cather. "In the food world, we know that not all meat is created equal. So for our family, we've decided we would rather eat less meat but make sure that what we are eating is grown with care for the animals, the people, and the planet. We definitely try to eat plant-based food as much as possible. It's not always easy with kids, but when I talk to my kids about why we're doing this, they're much more invested in making those changes than if I say, 'Oh, we're not going to have a hamburger.' If I tell them the reason behind it, they're empathetic people and they can understand."

We eat very little meat, but we do eat a lot of fish. We live in a commercial fishing town. One of my daughter's favorite playgrounds sits across from the fishing docks. After an hour of climbing, sliding, and swinging, she and I often watch the boats go in and out of the harbor, and we stop to buy scallops from the fishermen who harvested them or kelp from the seaweed farmers before heading home. Those interactions are setting the foundation for my child's understanding of the food system.

Limit Ready-to-Eat Items

You know how the fresh, better-for-you food is typically on the outer perimeter of the supermarket? It's the same with better-for-the-planet food. The less processed food you consume—ready-to-eat items, snacks, or frozen meals—the lower your carbon footprint. And yes, sometimes you need to buy the Goldfish or the animal crackers or your kid's favorite road trip snack. That's okay, although if you find yourself buying one snack or one ready-to-eat or frozen meal every week, try to figure out a homemade alternative. A simple Google search will lead to plenty of recipe options. We enjoy making homemade tortilla chips from corn tortillas.

Limit Palm Oil Consumption

It's almost impossible to avoid palm oil, especially if you're buying packaged food items. Made from the fruit of the African oil palm tree, it's an efficient, versatile crop that is used as a preservative, for frying, and more.

Yet conventionally produced palm oil is also responsible for massive deforestation and biodiversity loss in the tropical nations that produce it, like Indonesia, Malaysia, Colombia. Forests, once alive with animals like orangutans and other wildlife, are being cleared to make way for palm oil production. These forests also are home to people whose houses are destroyed and whose communities are torn apart. If you want to really dig into the effects of the palm oil industry, read journalist Jocelyn Zuckerman's book *Planet Palm*.

As with beef, not all palm oil is created the same. Try to find products certified by the Roundtable on Sustainable Palm Oil, which ensures that companies producing and/ or using palm oil follow a set of environmental and social criteria.

How to Raise Climate-Resilient Eaters

Making climate-friendly food choices at the supermarket, farmers market, or wherever you shop for groceries helps to set the foundation for raising climate-resilient eaters, but it's just one step. Kids also need to know and understand where their food comes from, how it's produced, and the impact it has on the environment, the workers who make it, and the local economy.

"When we are talking to kids who are going to be leading the next generation, talking to them about how [the food] they eat and how it's grown will impact the world that they live in is a no-brainer," says Cather. "It's really up to us to make sure we're giving them the right information, [as well as] giving them tools for little things they can do in their daily life and [how to talk] to those around them to make changes."

Most of us are vastly disconnected from where our food comes from compared with our grandparents or parents. I grew up hearing stories from my mom about how she would watch her grandfather butcher chickens (the chickens ran around with their heads cut off for a few seconds). Between that and my parents' telling us that we were having "Mary's little friend" (lamb) for dinner on Easter, it's a wonder my siblings and I eat meat. But my parents never hid where our food came from; it wasn't simply from the supermarket. Food came from an animal on a farm, from the sea, or from a plant.

I think it's important to be up-front about the food you eat. Some kids will enjoy eating meat and ask what animal it is; others will not touch it, because they know it's an animal.

"I have one child who's decided he's a vegan. Great," says Briana Warner, the CEO of Maine-based kelp company Atlantic Sea Farms. "The other child has decided that he's good with meat. Great. I've made the decision that I'm still going to eat meat. That's a decision that's right for me, but I wanted my kids to decide what they wanted."

To get to those decisions, though, Warner had to explain where and how the food they eat is made.

"When my kids ask where their food is from, I don't just say it's from a chicken," says Warner. "I tell them where it's from, and sometimes I'm like, 'You know, I don't actually know, and that's a problem. It's from a chicken, but I actually have no idea where this chicken is from. It was [cheap], and it was on sale, and I didn't have a lot of time. But let me talk to you a little bit about why it's a problem that I don't know where it's from, and the food system, and how it's much cheaper for us to buy bad food.' I want to be honest and take responsibility but in a way that's not guilt laden."

Whether your child decides to be a vegan, a vegetarian, or an omnivore, give them agency over their food choices. Of course, teaching kids about what we eat is hard to do when most of our food comes prepackaged and might not resemble where it came from. Depending on where you live, you might not be able to visit a farm, but you can still grow food, even if it's simply basil on a windowsill, and you can take your kids shopping with you. Even the supermarket offers plenty of opportunities to talk about food.

"I did an interview a few months ago, and they wanted to know, 'How do you feed your kids?,' what my secrets are," says Warner. "I think they wanted me to say I'm vegan, [and that] I source everything locally, but I was like, 'I do the best I can, and sometimes that's a bag of cut-up produce from Trader Joe's.'"

On Labels

Ask people who know me, and even those who don't know me well, and they'll likely tell you I'm very organized. My brain likes to sort things into categories, whether it's weekly tasks on a to-do list, books on shelves, or things in basement storage. But this sense of organization fails at certain times—like in the case of supermarket shopping. I go to the store with a list, but I'm easily distracted, especially by all the labels on food and drink packaging. The words *Fair Trade Certified*, *organic*, and even *natural* will catch my eye, and I'll have to pick up the product to take a closer look.

The number of labels on food and drink packaging these days is overwhelming and stressful. Labels can be helpful, but they should never be the sole or even the primary reason you decide whether to buy something. The most important thing about them isn't so much what they're trying to tell you but that they can be a starting point for talking to kids about how [a certain food] is made and the effects of making that item on people and on our planet. It's also an opportunity to learn how we evaluate information. Do we trust what that label is telling us? Remember, many of these labels are simply a marketing tactic, especially as almost all require producers to pay a fee for certification. Teaching kids to be skeptical of information and how to determine if a source is reliable are essential tools not just for deciding what to eat but for adulting in general.

One of the reasons I love my farmers market, even though I can't do all my shopping there, is that I can ask questions about how that tomato is made. Also, by living in the same town where that tomato is grown, you know how the farmer produces it and how they treat their workers. People talk. And thanks to the internet, we all live in a small town when it comes to learning about where our food comes from. If I buy something frequently at the supermarket, such as leafy greens, yogurt, or a particular cheese, I research the brand. That is something you can have your kids do as well. Go to the brand's website, read its story, check out their social media feeds to see what people are saying about it, and google it to see what's being written about the brand. Again, remember to evaluate the source and determine if it's reliable. Once I find a brand that meets my standards for sustainability, nutrition, taste, and price standards, I stick with it.

On Packaging

Most of the food we buy at the supermarket (especially processed food) comes in some sort of single-use packaging, often plastic, paperboard, metal, or glass. Takeout food is also almost always packed in a single-use container. All that packaging—be it plastic, aluminum, or Styrofoam—harms the environment and contributes to the climate crisis. Producing the containers uses resources like water and petroleum, and produces greenhouse gases, wastewater, and even heavy metals and particulates. According to the EPA, food and packaging materials make up almost half of municipal solid waste. So what to do? Start with your shopping. You can even treat it as a game with your kids.

- Shop at farmers markets, zero-waste stores, and the bulk-food aisles of your traditional grocery store.

- Limit the amount of processed and ready-to-eat foods you buy. Concentrate on products from the perimeter of the store.

- Plan meals. When you know what you're going to make and have already checked what you need, you don't end up overbuying and wasting food.

- Buy fresh bread from a bakery or the super-market bakery department and take it home in a reusable bag. If the available bread is already packaged, you can go to the counter and ask if they have any from the oven that you can take. Usually they do, or they'll set some aside in the future.

- Always select individual units of fresh produce instead of pre-bagged quantities and use reus-able produce bags, or skip the bag altogether.

- Buy anything with a long shelf life in the biggest size you can.

- Always choose aluminum cans or glass bottles over plastic.

- When you dine out, bring your own container for leftovers.

- Make it a game with your kids: See how many items you can find that have no packaging or at least no plastic in their packaging.

- Is there something your kids love to eat that you can find only in plastic at your supermarket? Can you have your kids brainstorm solutions? Maybe it's buying less of that product, or finding something similar elsewhere. Perhaps it's even having older kids contact the company (or elected officials) and ask for different packaging.

On Food Waste

I notice the blue speckle on the kitchen floor out of the corner of my eye. It's a deep navy color that's out of place on the old slanted wooden planks. I attempt to ignore it. I write more and answer some emails; still, there's that blue spot. Eventually, I get up to investigate and discover a blueberry my daughter must have dropped and that the dog surprisingly didn't gulp up straightaway. Instead, it is smashed into the floor and hard to remove.

Kids have a well-earned reputation for being messy, especially when eating. But as we teach them to use forks, knives, and spoons, and to cook and clean up, we also need to teach them about food waste. According to the United Nations, a third of all food goes into the trash. It would be easy to think that the vast majority of this food waste happens in restaurants, grocery stores, or even on farms. But in reality, the single largest source of food waste is in our homes.

"It's really hard because all our kids waste food," says Cather. "[I tell them], if they don't finish a snack, it will be their snack after school—that kind of thing. It doesn't always work, but trying to implement those practices, and again, telling them the reasons behind it, I think makes it a lot more meaningful than just saying, 'Don't waste your food.'"

Cather explains how if her kids waste food, it goes to a landfill, sits there, and impacts the climate, but she doesn't stop there. She's focused on how she pushes their larger school community to make the cafeteria greener, such as by composting food waste and doing away with disposable utensils.

Food waste is responsible for at least 6 percent of global greenhouse gas emissions. And that doesn't account for the energy, water, packaging, and transportation required to produce the food in the first place. The good news is that for most individuals, reducing food waste is the best thing you can do to reduce carbon emissions. And it's possible to do. Between 2007 and 2018, the United Kingdom cut farm food waste by 15 percent through a combination of initiatives, including expanded curbside collection and education.

You can start reducing food waste habits by meal prepping, storing food properly, reusing leftovers, and composting. These habits will set the stage for how they think about reducing waste in other areas while helping to teach and reinforce collaboration, communication, creative problem-solving, critical thinking, and systems thinking skills in the process.

Tackling food waste begins with your trash. Get the whole household involved in taking a trash audit. Give everyone some rubber gloves and go through your garbage and recycling, looking for wasted food and food-related items. For instance, are there foods you throw out often because they've gone bad before you can consume them? Do you have a ton of plastic drink bottles or individual yogurt containers? Is there anything that could've been composted?

Kate Bratskeir, author of *A Pocket Guide to Sustainable Food Shopping*, details how to do a trash audit in her book and suggests looking at everything with an eye toward reducing the waste involved to produce it. It could mean buying bigger yogurt containers or making your own, or not buying as many bananas if you find yourself constantly tossing them out. Once you have a good idea of the types of things you are routinely throwing away, it's time to think about shopping.

Meal planning is a pain. It's my least favorite activity, but it reduces food waste. It doesn't have to be some big affair, though. Before you go to the store, take inventory of the food in your home and determine what you have to buy. Ask your kids to come up with at least one thing they want for dinner each week—making sure they include a protein, a vegetable, and a starch. Having someone else come up with at least one idea for a meal cuts down on some of the emotional labor of cooking. It also helps kids learn some nutrition skills and gives them practice with ethical eating and not wasting food. From there, have two or three other loose ideas for meals. Chances are you'll fill in the remaining days with leftovers, takeout, and so on throughout the week.

I also like to do one day per week of use-what-we-have, when we come up with an idea based on the existing food in the fridge and pantry, especially food that will likely go bad soon. On an individual level, skipping takeout when there's a fridge full of food or making something from the few ingredients we have helps reduce consumerism, household waste, and our carbon footprint. On a collective level, solving the climate crisis requires delayed gratification. It's cooperating now to mitigate damage even though we won't see the results.

With a bit of preparation and by storing food properly for your climate, you can reduce your food waste and ensure that when you reach into the fridge for that head of lettuce, it will still be fresh.

One of the beauties of "use-what-we-have night" is that it inspires creativity. It forces us to look at what we have and come up with something to make. The same can be said for any leftovers or food scraps. When meal prepping, think about how you can use every bit of food. For instance, those carrot tops, or other green scraps, can become a pesto.

My daughter is often in the kitchen when I'm cooking, either playing with some of her toys or coloring by the counter. It's easy for me to look up from what I'm doing and say, "Hey, what do you think we should do with this?" We take a similar approach to any random leftover food. If she doesn't finish all of a snack, for instance, we save it for later. The same goes for a meal. Chicken breast or salmon that we didn't manage to finish gets saved and becomes the start of a grain bowl or tacos combined with any leftover vegetables we have on hand.

As we're clearing the table, food waste is another thing I talk through with my daughter. I'll reiterate why we're trying to waste less food and then ask her to help me develop solutions. We'll count how much food is left over (this is great for helping younger kids understand what reducing waste means), think about what we can do with it, and decide what we should make less of next time. I'm trying to get her to think about valuing food and spark her problem-solving skills by coming up with other uses for it.

In addition to helping the planet, wasting less food will benefit your wallet in the long run. The average American family of four throws out $1,600 a year just in produce.

On Composting

Composting with kids helps introduce them to sustainable living—and may make the process feel like less of a chore for you while reinforcing skills about contributing to a household and community and helping them practice delayed gratification.

The beauty of composting is that it can be as simple or complicated as your child is ready for. If your kids are young and just learning about cleaning up, try to seamlessly integrate it into your cooking and after-meal clean-up routines. For instance, have your children sort their breakfast, snack, lunch, and dinner into the trash, recycling, and compost. Pick just one meal a day to start with. Once your child has mastered the basics of trash, compost, and recycling piles, you can begin introducing composting and how it works.

At first, you can make a game out of it. When you take a walk, for instance, point at different things and ask if your child thinks they could become soil. As your kids get older, you can have them experiment. For example, you can have them cut up food scraps with a designated pair of compost scissors and explain how, just like we need to chew our food, compost needs small pieces to work, and talk more about organic versus nonorganic material. Let them see which items can decompose by putting something compostable in one container and noncompostable materials in another. Then let them observe the containers daily so they will know the process of decomposing.

The older they get, the more you can explain why, when we put organic waste in the trash, it doesn't break down as it would in a compost pile, because it is buried under other debris, rotting, and emitting the greenhouse gas methane.

It takes time. In about a year, you'll get soil you can use in your garden, making the whole process a lesson in delayed gratification. With the soil you get from your composter, you can teach your children how to grow food for themselves. You can grow blueberries or kale, taking the soil from your composter and using it in your garden.

Watching the waste decompose will fuel their curiosity, leading them to want to compost more and learn more about the world around them, or at the very least, it'll become an ingrained habit.

While reducing food waste can make a significant difference in your household carbon footprint, it can be a jumping-off point for how kids (and all of us) think about other types of waste and our overall consumption of resources.

COMPOSTING BASICS

What is compost? Simply put, compost is the mixture of decayed matter that results from organic material—such as food scraps and yard waste—breaking down, with some help from microorganisms, into beautiful, healthy soil that gardens love.

While what exactly you can compost will vary, especially if you are one of the few people who have curbside compost collection, you can generally compost anything that is an organic material and won't attract pests.

Do compost leaves and other yard trimmings, fruits and veggies, eggshells, coffee grounds, nutshells, rice, and other grains.

Don't compost meat, fish, dairy, oil, pet waste, or yard waste treated with chemical pesticides. An exception to this is if you have a Bokashi composting system that you can use to compost meat, fish, and dairy.

Alternatives to backyard compost: If you don't have a yard, there may still be ways to compost. A growing number of municipalities offer composting, and that number is expected to grow. Give your local solid waste department a call (you can normally reach it through your town or city hall). Some community organizations, especially gardening clubs or farmers markets, offer drop-off options for compost. There are also companies that offer subscription curbside composting services outside. Although there's no one directory of those services, organizations like the Institute for Local Self-Reliance or, again, your local government, can point you in the right direction. There are also some subscription services that allow you to mail back countertop compost, and that turn the food scraps into animal feed. While there may not be an accessible, affordable option in your area, working to reduce your overall waste is still a way to practice problem-solving with children and help connect conversations about the environment with action.

Composting systems: There is an overwhelming number of compost bins and systems on the market. The best is the one that's going to work for you and your household. A dear friend who lives in New York City struggled with composting for years for various reasons. One day, her partner hung a plastic bag from one of their kitchen drawer handles to collect scraps while he prepped dinner, and to her surprise, it became the very unsophisticated compost bin she needed to get into the habit.

Whichever system you decide on, make sure you consider both the size and the location. You want your compost bin to be fairly accessible and big enough to hold everything you need it to; otherwise, it'll be tempting to throw things into the rubbish bin.

Layering: Alternating layers of food scraps and yard waste will help to speed up the composting process.

Turning: Turning your compost will also help to speed up the composting process, but as long as you have the right mix of organic materials, you'll still get compost; it will just take longer.

Watering: Watering your compost is an essential part of the process. You want the pile to stay moist, because it helps with decomposition and keeps the pile's temperature at the right level. Ideally, you should water it once or twice a week.

8
Consumerism

We have the knowledge, the technology, and the money to prevent the most dire climate scenarios from becoming realities. So why haven't we done this? How do we stop the momentum of the systems that contribute to the climate crisis?

Tom Bailey is well versed in climate-change issues. The London-based engineer has a background in green energy and sustainable buildings. Yet when he was head of research at C40, a global network of mayors working to help limit global warming to 1.5°C/2.7°F, they did research that challenged everything he thought he knew about mitigating the climate crisis.

"It pulled the rug out from under my feet," Bailey recalls. "I'd spent my whole career working on new technologies, new policies, things to help reach net zero, including wind turbines, energy-efficient buildings, mass transit transport infrastructure, but this research looked at all the trade that goes on globally, looked at where we are today and how we get to net zero. It basically showed that no matter how much you try to deliver on the green solutions, policies, and technologies, you can't green fast enough to keep up and reach net zero, as long as our society remains focused on more and more stuff. The rate of consumption just keeps growing at such a rate that you can't keep up with it."

There's little doubt that Western countries where we consume the most will need to dramatically shift their culture of consumption to best handle the climate crisis. Even the most eco-conscious among us often live in ways contrary to what's best for the environment. And often we do so without having a choice. Yet individual actions, even imperfect ones, are essential to adapting our culture to the climate crisis, and taking action is part of learning resilience, even if it doesn't always turn out the way we want. It gives our children and us a powerful feeling of agency.

"As long as our society, our culture, our mindsets, our systems, our education systems, our political systems, our economic systems are focused on more and more and more stuff, you just can't get in balance with nature," Bailey says. "It really shook me, because it showed that everything I'd been working on my whole career, while it was still really important, wouldn't be enough."

While the onus to solve the climate crisis lies on governments and corporations, a report published in 2022 titled *The Future of Urban Consumption in a 1.5°C World* found that individuals can reduce greenhouse gas emissions by up to 27 percent. How? By making six significant shifts in their day-to-day lives:

1. **Eat a primarily plant-based diet and minimize food waste.** Remember, plant-based diets yield significantly less heat-trapping gas than diets rich in meat.

2. **Buy no more than three new items of clothing per year.** The fashion industry accounts for up to 8 percent of global carbon dioxide emissions, according to the UN Environment Programme. That's more than the emissions from worldwide shipping and air travel combined.

3. **Keep electronics for at least seven years.** The study found that the emissions associated with a single Apple iPhone 11 Pro result primarily from its production, transort, and end-of-life processing (86 percent) and not from the actual usage of the phone (13 percent).

4. **Keep flying to a minimum (no more than one short flight every three years and one long-haul flight every eight years).** Remember, aviation currently accounts for around 2 percent of overall global emissions, but those emissions are growing faster than those from any other form of transportation.

5. **Ditch your car or keep your existing vehicle for as long as possible.** Transportation is responsible for about a quarter of greenhouse gas emissions, and while air travel is an easy climate target, more than two-thirds of transportation emissions come from vehicles on the road. Their use also contributes to air pollution. And while electric vehicles are slowly making inroads and release significantly less emissions than gas-powered vehicles, they still contribute to pollution.

6. Make at least one shift to nudge the system, like divesting from fossil fuels, switching to green energy, or getting your child's school to do so.

After processing the research, Bailey decided to do something about it and helped to found Take the Jump. The organization aims to help people make these jumps while living joyfully, with workshops, events, tool kits, and a community network for support and inspiration.

"We are very far from powerless. Our action is meaningful and impactful," Bailey tells me. "If you look at any of the big society-wide revolutions of history, the Industrial Revolution, the scientific revolution, civil rights, gay rights, gender rights, none of these things started with a leader who said, 'Oh, this is a good idea,' and then told everyone else how to do it. They started at the level of normal people deciding the world needed to be different; they started to think differently, shifting their own communities and cultures. That's what led to system change. Demanding a new economic system won't happen until we have a real cultural example of what the alternative is and the drive for it."

How do we raise children who do that? Personally, I look at that list and feel overwhelmed. How can we adopt some of these practices while still living with joy?

We need a cultural shift that moves us away from an instant-gratification society. Studies show that people who have a higher tolerance for delayed gratification thrive more in their careers, relationships, health, and finances. Still, developing a tolerance for delayed gratification is no easy task in a world where we've gotten accustomed to same-day delivery services, getting groceries and goods drop-shipped from across the globe, calling an Uber with the push of a button, or streaming entire seasons of TV shows within days.

What do you do when you're having a bad day or feeling low? You're not alone if you get out your phone and buy something.

"You get the endorphin high, but then after ten minutes it's gone," Bailey says.

How You Can Make Those Shifts in Your Own Life

Start small. Go back to the trash. By now, your kids somewhat understand sorting and have looked through it with an eye toward reducing food waste and waste from food packaging, like those yogurt containers. But how is your family actually doing? Pay close attention to the plastic you discard. It's estimated that between now and 2050, the manufacturing of plastic could create fifty times as much pollution as every coal power plant in the US. Give your kids some rubber gloves and have them look through the trash for items that could be removed. Start with the food-related items as a check-in on how you're doing and make a plan to adjust anything you might need to; for instance, if you're still wasting those bananas that are about to go bad, try peeling and freezing them for smoothies or other desserts. After you've considered the food-related waste, check in on other items. Is there anything that could have been repurposed? What about recycled? What about things you just didn't need to buy? Ask your kids.

In 1960, according to the EPA, 82.5 million tons of rubbish ended up in American landfills each year; by 2018, the last year for which data is available, that number had grown to 146.1 million tons, or 2.4 pounds per person per day. That number also doesn't include trash that is recycled, composted, or burned; made into animal feed or bio-based materials; processed biochemicals, co-digestion, or anaerobic digestion; donated; or used in land applications or sewer or wastewater treatment. All in all, about 50 percent of US waste ends up in landfills.

Landfills are, of course, a necessity. They help prevent contaminants from ending up in the environment, but they're also a huge source of environmental problems. The creation of landfills requires habitat and their emissions can result in various health problems for the people who live near them, who are often low-income and minorities. They can also release toxins into the soil and groundwater. As rubbish breaks down, it releases greenhouse gases, mainly carbon dioxide and methane. Landfill sites contribute roughly 20 percent of the global anthropogenic methane emissions. While methane makes up only about 20 to 30 percent of global greenhouse gas emissions and sticks around in the Earth's atmosphere for only about a decade, it's estimated that Methane warms our atmosphere twenty-five to thirty-five times as much as CO_2 does. In other words, cutting methane gas emissions quickly could give us time to find longer-term solutions to reducing carbon dioxide emissions.

Older kids who are already aware of the climate crisis and learning about greenhouse gas emissions will grasp the concept of reducing waste to help the environment.

"When it's time to give someone a gift, my eight-year-old will go through our recyclables and she'll find something she can use to wrap it," says New York City mom Chantal Alison-Konteh about her daughter's efforts to reduce waste. "I never told her to do that; it's her own thing, and it really matters to her."

Reducing waste may be more challenging with young kids. If you've already gone through the composting process, though, you can start talking about how our little bit of waste added to everyone else's little bit of garbage creates problems for the planet, so we want to do what we can to reduce the amount of waste we have.

Children tend to be highly empathetic. This isn't to say they can't be a bit lazy or selfish at times, but they generally care about the world around them, and it's up to us to give them gentle reminders to reduce, reuse, and recycle. But they also need to see the impact of trash and have adults in their lives who are also working to reduce waste.

To put these things into perspective, it helps to get outside. I'm lucky enough to live by the coast, and my daughter and I often find ourselves at the beach no matter the season. These days, we take three pails with us: one for collecting beach treasures, one for collecting some seaweed to add to the compost or to press into art projects, and one to collect any bits of waste, mostly plastic, that we find. We talk about where we think the waste comes from. Seeing waste helps kids understand the larger concept of why trash is harmful to the environment, but it also serves as a way to continually motivate them and ourselves to reduce our waste at home.

"Parents should teach their kids the impact of small choices that they can make every day, so kids will be enlightened to the fact that repeatedly choosing positive choices becomes ingrained and habitual," says teenager Clare Flaherty, adding that constantly electing to drink from a glass instead of a plastic cup at a young age made it easier to always do that as she got older. "Parents should be very clear that we all have a responsibility to take care of this land."

It helps if they see you doing this. The goals when it comes to reducing overall waste are similar to those for food waste: Focus on not purchasing things you don't need, try not to buy stuff in wasteful packaging, reuse or give away what you can, and recycle anything else.

"I really believe individuals can make an impact," says Brooke Petry, the South Philadelphia mom. "I believe it's important because I think you're getting that muscle memory for considering your impact and considering the impacts of things on the planet, on future generations, and the more you're in that mindset, the more able you are to advocate for the kind of systemic change that we need."

Recycling comes with its own set of problems. It puts the responsibility for environmental protection on individuals rather than on the companies creating the problem. According to the UN, more than 8.3 billion tons of plastic have been produced since the early 1950s, most of it made within the last two decades. Only 9 percent of all plastic waste ever produced has been recycled, and at least 60 percent has ended up in landfills or in the environment. Companies creating single-use plastics, from water bottles to laundry detergent containers to shampoo bottles, could use aluminum or glass instead. Yet they don't. Despite knowing for decades that plastic recycling wasn't a feasible end-of-use solution, they put the onus on consumers, encouraging us to recycle to solve the problem they're creating.

Our recycling programs are largely broken. What can and cannot be recycled often varies significantly between countries, states, and even towns. My parents live a mile from me, but in a different town. They can recycle some things that I can't, although until very recently, they had to sort their recycling into paper and plastic, whereas mine has been single-stream for ages. This is hardly unusual in the US, and it leads to a lot of confusion. Only about 35 percent of Americans actually recycle, even though 94 percent support recycling, and 74 percent say it should be a top priority. Not everything that ends up in the recycling bin gets recycled, for a variety of reasons, including containment, improper handling, and lack of financial incentives. Anywhere from 15 to 30 percent of recycled items still end up in the landfill. And recycling an old item and turning it into something new is a process that still uses up resources.

Teaching our kids to focus first on buying less and saying no to plastic, and then on reusing what they can, will set the foundation for them to say no to the consumerism that's harming our planet, and help them become creative problem-solvers. Encourage them to reach out to brands they're curious about and ask questions about where the product is made, whether the factory provides a living wage, what it does with waste, and what it does with returns. Businesses that care are happy to talk about their sustainability measures.

"I spend a lot of time thinking about it for my kids," says Jason Hine, an owner of a store that sells zero-waste products. "They should be worried about grades and friends and sports, not the climate crisis, but I like to think I nudge gently. I say, 'Hey, look at this, we're going to use shampoo bars instead of plastic shampoo bottles,' and hope that they then have those casual conversations with their friends."

You also have to keep having those conversations around environmental values, especially with teenagers. Michele Bigley, the regenerative travel specialist, tells me about being surprised to see a plastic squeezy water bottle in her house.

"[My son Kai] got a gift certificate for his birthday to a sporting goods store, and that's one of the things he bought with it," she says.

Their conversation went like this:

Michele: There's a reason you don't see these kinds of bottles in our house.

Kai: Yes, but those Klean Kanteen and Hydro Flask ones are so hard to open when you're playing soccer.

Michele: But there's a reason. There is all kinds of crap put in this plastic. It's not good for you, and it's not good for the environment. There's a reason why you don't see them in our house.

Kai: Oh, okay.

Michele: You already bought it and you have to make a choice about how to use it.

"Sometimes it's remembering to restate some of those lessons," Bigley says. "They might see it, but maybe I never really overtly said, 'We're not gonna buy crappy water bottles because of the PVCs.'"

When my child or I want something, I try to think, *Is this good for us, for the planet, for the people who made it?* It's easier said than done. I've been on a quest to rid us of fast fashion. The number of clothing items produced each year doubled between 2000 and 2014, according to the consulting firm McKinsey. A 2016 World Economic Forum report estimated that 150 billion new garments are created each year. Most of those clothes—about 84 percent—end up in landfills or incinerators. In the US alone, more than 11 million tons of textile waste ended up in landfills in 2018, according to the EPA's most recent data. And it's not just harmful to the environment: All that clothing production requires cheap, mostly unregulated labor. It's estimated that 98 percent of fast-fashion factory workers don't make a living wage.

Over the past several years, I have limited my new clothing purchases to used clothes and a handful of brands (fewer than five). These brands are B Corp certified (which means that the companies have met certain standards for social and environmental performance) or meet my own environmental standards, come from a shop I can walk to (and even those I'm slowly phasing out). It's harder with my child. I've had a lot of success finding used outdoor clothing and gear, from boots to jackets, and I get the occasional hand-me-downs from friends and family. But she also grows out of things quickly, and all that outdoor play is hard on clothes. We sew or patch some things, but in our town that kind of clothing repair isn't the norm, isn't cool, and I worry about her not fitting in when she goes to school.

It's the same with birthday gifts at children's parties. I look at the table full of presents and wonder, *What are we doing?* What values are we passing on? Because of the Covid pandemic, my child was four when she had her first birthday party with friends her own age. I tried to get her used to the idea that her friends' presence at the party was a gift, and I really wanted to put "no gifts" on the invitations, but I lost that discussion with her dad. The compromise was putting "no gifts necessary" on the invitations, but 99 percent of the guests showed up carrying brightly wrapped boxes or holding bags with tissue paper sticking out. They were lovely and thoughtful, and two of them included gift cards to local bakeries, an idea I have stolen as my go-to gift for children's birthday parties, but still, I looked at all the *things* my child received and wondered about the message she was receiving.

That we can't even come together to celebrate without needing to buy things is a glaring example of our consumer culture.

The vast majority of toys on the market are plastic. They're shiny and inexpensive, nearly impossible to recycle, and get tossed away faster than items made out of, say, wood, which tend to be longer-lasting and easier to pass on.

"It doesn't have to be that way," Bailey tells me when I mention birthdays. "It hasn't been that way for most of history. It's just a recent thing, you know."

He's right about that. If you look into the history of birthday celebrations in the US, they didn't really become an American tradition until the early twentieth century. Do you remember any of your childhood birthday presents? I can remember a few: the real puppy, for instance, that I got for my fifth birthday, and a giant teddy bear that I passed on to my child. Yet mostly what I remember from childhood birthdays is the feeling of being surrounded by family and friends and the time we spent together. I want to create traditions for my child around holidays and other celebrations that don't center consumption.

"We need a bit of realignment, but it's not going back to living in caves," says Bailey. "You don't have to give up everything you enjoy in your life. You can still go see the world. You can still eat. You can still look great. Just repair clothes, get them secondhand, recycle them, share them, swap them. It's about a balance." We need to change the way we consume, starting with a buy-less mindset, renting or swapping items whenever possible, finding new homes for things we don't want anymore, and treating recycling and trash as a last resort.

And we have to talk about what we're doing. Remember from chapter 1, "Talk," how sharing your reasons is important? This is how cultural shifts start.

"I'm not buying the latest trend," says Alexina Cather, mom of three and the director of policy and special projects at Wellness in the Schools. "And my kids know that. They go to the toy store, and they see [that a toy is], Oh, it's made in China. Or it's made of plastic or something flammable, and they say, 'I'm not even going to ask for that.' Between our values and now with my husband also being a fireman, the amount of things that don't come into our house is a lot. It's hard to communicate that to our family without them rolling their eyes. But it is what it is."

For the good of future generations we need to get more comfortable with doing things differently now even though we might not see the benefits.

Part 4 Act

We know how to talk about the climate crisis and how to make a difference in our everyday lives, yet we cannot mitigate the worst of the climate crisis without systemic change. This final section of the book explores how individuals can help nudge the system, starting with the energy system.

Your child's school can be a great place to start. School offers a built-in community that can be organized toward meaningful action. Building on curiosity, talking, and problem-solving, we'll explore how children and adults can create strong communities beyond their schools.

The final chapter looks at what it means to be an activist, and at how activism gives people a sense of purpose, a community, and a supportive network. Activism is a powerful tool for building resilience while fighting for change.

9 STEM

Large swaths of my daughter's day are a mystery to me. I know where she is, of course—at school—and thanks to emails from her teachers, I even see pictures and learn about the activities she's doing, but there's much I don't know. Once our children enter school, their world expands in marvelous ways. They're learning without us, and we no longer get to see or hear everything that's sparking their curiosity or how they're problem-solving who gets to line up first or where they fit in the ecosystem of the school.

When we're walking through town, at the farmers market, or just running errands, my child will now run into someone she knows. She's become a part of her own little community, thanks to school.

School quickly becomes the center of our children's lives, and it's also a place where they get to practice many skills and build the knowledge they'll need to move mindfully through the world, and where they can advocate for and see climate action catch on. Much of this begins with STEM education.

What Is STEM?

STEM, an acronym for "science, technology, engineering, and math," has been a cornerstone of education initiatives since the early 2000s, mainly because those areas make up popular and fast-growing industries. While it's impossible to know what careers children will have decades from now, STEM learning emphasizes skills—such as innovation, problem-solving, and critical thinking—that will be crucial as they grow into adulthood in a warming world.

"We were not just sitting in a classroom," says Tamar Shoshan, who attended Manhattan Hunter Science High School in New York City, which includes vertical farming as part of its STEM curriculum. "We did social-emotional learning and implementation, restorative justice practice. All of these things taught us how to tune in to ourselves, resolve conflicts, deal with other people, and care about the world around us."

STEM, sometimes called STEAM, focuses on natural, physical, and life sciences; technology-related disciplines, including computers and electronics; engineering; and math. The educators who prefer the term *STEAM* posit that the *A* focuses on the arts, including writing and reading. (Unless otherwise noted, I'm going to refer to STEAM and STEM interchangeably.)

STEM differs from traditional education programs because it takes an interdisciplinary approach to learning, but if you're familiar with the Waldorf or Montessori approaches to education, you'll see many similarities in the way lessons are connected throughout disciplines. Instead of teaching the subjects in isolated blocks, STEM programs teach all the disciplines in an integrated way. The goal is to connect what the kids are learning to real-life issues so they will be better equipped to tackle the challenges facing society.

"One of our school projects was on school foods," Shoshan says. "We had to create an alternative menu for school, and it was challenging because there are all these requirements from New York State and the US Department of Agriculture. We had to choose foods from the five different food groups, and they had to be under $1.60 per student. And then we also spoke about food insecurity and inequality of access. I learned a lot."

STEM focuses on how we think, asking students to verify and evaluate information and use their critical and analytical skills to solve a problem.

STEM programs look different in different schools. For instance, while Shoshan's school focuses on food justice, others focus on health care or renewable energy. Given her late-fall birthday, my child was privileged to have the choice either to start kindergarten at four at the public school in our town or, thanks in large part to the childcare funding benefit my editing job provides, stay in a private nursery setting for another year before starting primary school at five. As such, I went through the exhausting process of touring different schools and considering the options. There was the private school with a full-day prekindergarten that was very STEM focused, and the Montessori school, which focused less on traditional sciences but used taking care of chickens and two llamas as part of its lesson plans. I know that having an early connection to animals can help children practice nurturance, an essential skill for mitigating the climate crisis. There was also the option of keeping my child at the nature pre-school she had attended for the previous two years, which was centered on the acres of trails on the property. As I considered the best option for my child, who, at age four, I thought would have a hard time sitting still for six hours in a traditional school environment, I was also aware of the benefits of an interdisciplinary approach to her education.

The key benefits for all students, even those who won't necessarily have a traditional STEM career, are that these programs foster critical thinking skills, says Bejanae Kareem, an education consultant with more than twenty years of experience in pre-K–12 and postsecondary education, grants, equity building, and STEAM education.

"They're life skills: being able to communicate, critical thinking, and creativity," says Kareem. "With STEM-integrated learning, students can see the connections between this and that. It's not 'Oh, I don't like science, science is something over here'; it's 'I am science and I see it throughout my day.'"

For instance, a school district with edible gardens might have elementary students learn the gardening process: What is a seed, what happens when you plant it, when you water it, and when you don't? Middle schoolers might learn how the process works and have to come up with solutions for drought, and high schoolers might take the process even further by teaching others how to garden. That garden immerses students in everything from engineering, in designing and building raised garden beds; to financial literacy, especially if they sell the fruits and vegetables at a local farmers market; to the mathematics of yield.

STEAM teachers are also different from traditional instructors. Their goal is to serve as facilitators, helping children learn the process of thinking and assisting students in finding solutions to problems instead of instructing them on the correct answer. Teachers learn from students just as much as they teach them—something parents can practice too.

If your child asks you the meaning of a word, do you simply tell them, or do you ask them how they could figure it out, guiding them toward finding context clues, looking it up in a dictionary, or, I guess, asking Alexa so they can discover the answer on their own?

With STEM, there is often more than one correct answer, because the goal is to solve an issue. Students are learning how to find solutions, and often there are multiple solutions. Like with the climate crisis, there are numerous ways to cut fossil fuels. In some communities, this means solar panels; in others, it means wind turbines.

STEM Skills Are Climate Skills

STEM classes let students nurture their creativity, practice their problem-solving skills, and even fail on the quest to find a solution. STEM lessons also often focus on teamwork, which is essential for solving problems like where to place a garden bed or deal with drought, but also let them practice building community by building interpersonal and collaboration skills.

"Some schools pick a topic that makes sense for their community, so in Alaska that might be renewable energy using wind turbines," says Kareem. "Here in Atlanta with the daylight hours we might use solar panels. At many schools I've seen urban gardening pop up, especially at schools that are located in food deserts, where the community doesn't have much access to fresh food."

Schools and Fossil Fuels

"If you care about justice and our children's future, I am appealing directly to you: Demand that renewable energy is introduced now—at speed and at scale; demand an end to coal-fired power; demand an end to all fossil fuel subsidies." Those were the words of UN Secretary-General António Guterres upon the release of the IPCC's April 2022 report *AR6 Climate Change 2022: Mitigation of Climate Change*.

Look around. Can you find anything within eyesight *not* produced with fossil fuels in some way? I can see my stovetop from where I am sitting. It's electric, and I have solar panels on my house. Yet I'm almost positive the stove was delivered using transportation powered by fossil fuels and was probably made at a production facility also powered by fossil fuels. And that's just one item in my view. The curtains on the windows, the dishcloth on the counter, the dishes in my kitchen cabinets, the food in my fridge—fossil fuels touched them all. The only thing I can find that wasn't made in some way from fossil fuels is the seashell sitting on the window ledge that my daughter found at the beach.

Fossil fuels are ingrained into every aspect of our daily lives. We can use all the reusable canvas totes we want and teach our children to reduce waste and consume less (and make no mistake, those actions are critical too). At the end of the day; however, we cannot protect our children and grandchildren from the worst-case scenarios of global warming unless we divest from fossil fuels, starting with the energy system.

It will not be easy to divest our lives of fossil fuels. (Did you find anything within your view that fossil fuels had not touched in some way?) The fossil fuel industry is powerful. It has long sent more delegates than any nation has to the annual UN Climate Change Conferences. And industry leaders have known since at least the 1950s that the creation of carbon dioxide from burning conventional fuels such as oil could be catastrophic for the planet. Yet they've continued to push for the burning of more oil, gas, and coal.

According to a 2017 study published in the journal *Climatic Change*, of the ninety companies worldwide responsible for nearly two-thirds of greenhouse gas emissions from 1880 to 2010, eighty-three produced coal, oil, and/or natural gas. And while the International Energy Agency (IEA) predicts that emissions of carbon dioxide will reach a peak by the mid-2020s and then drop slowly in the decades after that, that prediction still puts us on track for the world to heat up by 2.6°C (4.7°F) by 2100 from pre-industrial levels—an increase the United Nations says would be "catastrophic."

Remember that under that scenario, according to the IPCC's August 2021 report *AR6 Climate Change 2021: The Physical Science Basis*, the Arctic Ocean would be ice-free in the summer and there would be a significant drop in global food production because of widespread crop failures. Children alive today could see entire ecosystems collapse within their lifetimes.

To avoid that, we need to act now. The world needs to move away from fossil fuels much faster than it's currently on track to do, in order to hold global warming to as close to 1.5°C as possible, a scenario in which children today will still face more extreme weather but not as many dire consequences.

According to the IEA, which in 2021 laid out a roadmap of what would be required to hold global warming to 1.5°C from an energy standpoint, wealthy countries would need to shut down virtually all fossil fuel power plants by 2035, transitioning to using wind, solar, or nuclear power.

The world can do that. We can do that. Many students and parents already are. In 2013, eighth graders in Lowell, Indiana, asked their school board to explore alternative energy options. The Tri-Creek school district responded by installing more than 100 solar panels, which in the twelve months preceding August 2021 saved the district $140,000 in energy bills.

That school district is just one of a growing number around the US and Europe that are using renewable energy to save money and power schools with clean energy while providing children with incredible hands-on learning. They give me hope that individuals working together can help decarbonize their communities and that schools within my child's own district will also be able to convert to renewable energy.

Drive through a residential neighborhood these days, and chances are you'll see at least one home, if not several, with solar panels adorning the roofs. As the cost of renewable energy, especially solar, has declined, more and more homeowners have added panels to their properties.

According to the Solar Energy Industries Association, the cost to install solar has dropped by more than 70 percent over the past decade.

Renewables made up nearly 21 percent of utility-scale US electricity generation in 2022, the last year for which data is available. The bulk comes from wind power (10.3 percent) and hydropower (6 percent). Solar generation (including distributed), which powers roughly 4 percent of homes in the US, is the fastest-growing electricity source in the country. Meanwhile, the proportion of electricity generated by renewables in the UK has grown to more than 40 percent, outpacing that from fossil fuels. However, the UK does include bioenergy (12 percent) in its definition of renewable energy. Bioenergy involves burning wood pellets, which has more negative climate implications than wind or solar power. Overall, renewables accounted for 38 percent of EU energy sources in 2020, compared with 37 percent from fossil fuels. According to the European Commission's 2021 *State of the Energy Union Report*, this was the first time renewables overtook fossil fuels as the number one power source in the EU.

Yet renewables are still out of reach for many individuals and families. When I bought my house in 2016, it was a bit of a disaster. (It's still a bit of a disaster, to be honest.) At the time, no one had lived in it for nine months, and there was a long list of things that needed to happen to make it comfortably inhabitable. Yet one of the first things I did was install solar panels. I wasn't a mom yet, but I was betting on the future. While it was relatively inexpensive for me to add the solar panels, I still ran into several problems. Because my house had been vacant for so long, there was no way to accurately determine how many panels I needed. More than six years later, and now that I have an electric vehicle, I need to add more, but because of several factors, including changes to some of the solar options, it's not feasible for me to do that right now.

What am I doing instead? The remaining energy, which I get from the local utility company, has an option for me to select a renewable energy supplier, which I've done. But not all US states allow consumers, especially residential consumers, to choose their energy supplier, and some that do either have higher costs for renewables or don't allow for 100 percent renewables. Still, it's a good option for both homeowners and renters to choose if they can. But if we really want to make an impact on our energy sources, we should be looking at schools.

A Chance to Teach Change

"Schools are often the single largest energy consumer in their area, so if we can make a dent right where our children spend their waking hours five days a week, it will help them understand they can make an impact in a hands-on way while also giving them a lesson in STEM," says Lisa Hoyos, the national climate-strategy director at the League of Conservation Voters. In a previous role at the Sierra Club, Hoyos began a program that works with children and their parents to encourage schools to move toward clean energy. The organization provides tool kits that include information on overcoming specific challenges (including strapped school finances), the benefits of clean energy, and how to work directly with parents and school officials looking to make the change.

According to the Sierra Club, schools in the US spend nearly $8 billion per year on energy costs; it's the second-largest expense most schools incur. The first is personnel. According to the Carbon Trust, the numbers are similar in the UK, where schools spend more than $630 million annually on energy costs.

Those numbers mean that many schools have a tremendous opportunity to save money by switching to clean energy. In the US in 2021, the last year for which data is available, 9 percent of K–12 public and private schools used solar power. There's also a small yet growing number of schools—estimated to be around 100—that use wind energy. It's not just money that school districts save, however. Renewable energy is cleaner, and switching to it means schools are helping to protect the health of their own community members if they have energy plants within the district, or the health of people in other towns and communities, especially children, who are often more vulnerable to air pollution. According to the World Health Organization, every day around 93 percent of children under the age of fifteen breathe air so polluted, it puts their health and development at serious risk.

How STEM Meets Renewables

Schools that use renewable energy often incorporate STEM learning into their curriculum. Science teachers, for instance, teach about where power comes from and how it's used, health teachers teach kids about the effects of clean energy on air quality, and student health and social studies classes have learned about the politics of clean energy. While my daughter's future school district does not have solar panels, we talk about ours at home. She knows what they do and, in simple terms, that they are better for the planet, and that we are able to choose to get our electricity in a way that benefits other humans, animals, and plants.

While she's too young to fully comprehend the idea of electricity, she is starting to understand and appreciate the power of the sun. At the beach near our home, we collect seashells, dried seaweed, and beach grass. After rinsing ourselves and the seashells, she places the dried seaweed and beach grass on sun art paper in a shady part of our yard. These papers are pretty inexpensive, and after she's set up her objects on the pieces of paper, we carefully place them in our sunny walkway. We let them sit for five minutes before placing them in a tub of water for a minute and then laying them out to dry, after which she gets to see how the sun's energy made art.

Ideally, by the time she's at school, it will have made the switch to clean energy and even made it a part of student projects, as other schools are doing.

For example, the nonprofit Renewable Energy Alaska Project (REAP) educates K–12 children on clean energy through the lens of STEM.

"One of the goals is energy literacy and making sure people understand where energy comes from, and it's way easier to teach a kid," says Greg Stiegel, deputy director of REAP.

Established in 2004, REAP helps schools transition to clean energy, which tends to be wind or hydropower in Alaska. The state, which traditionally leans red, has many communities on the front lines of the climate crisis. Melting permafrost and coastal erosion mean entire villages in the state will need to be relocated, such as Newtok, Alaska, which was relocated in 2019 at the cost of more than $100 million. Looking at energy through the lens of STEM, Stiegel tells me, has been a way for the nonprofit to talk about the climate crisis in a less polarized way.

"Energy is expensive here; you can't argue around that," Stiegel says. "Clean wind and solar are the cheapest on the planet, and it's an easy connection to make."

While REAP brings STEM classes on renewable energy to many different schools, it has found it works the best in some of the more remote communities, where REAP puts wind turbines on school properties, so the children can see them firsthand.

For example, some schools have installed wind turbines in their courtyards or have energy-use tracking mechanisms in their lobbies as tools for classroom learning. Antelope Valley Union High School District in California integrated its solar power system into the science and math curriculum using an online energy monitoring system. Students scored 60 percent higher on tests in these subjects after solar-focused lessons.

REAP also works with vocational schools and colleges to help interested students transition to clean energy careers, yet it knows not everyone will go into those jobs.

"[People] who can understand energy issues on a deeper level are going to be better stewards of it," Stiegel says. "In five, ten, twenty years, when these children own homes, it will help them be more engaged and understand efficiency. It's really about educating them on where energy comes from and bringing in the climate piece, decisions you can make, lesson plans, doing mini energy audits of their houses. And then when they grow up, they understand the importance of energy-efficient homes. It's potentially a pretty powerful tool for change."

Raise your hand if you knew very little about energy before renting your first apartment or buying your first home—*me, me.* I was in my mid-twenties before I really understood energy and the impacts of my choices on other people, communities, and the planet. This is something I want my child to grow up knowing. Not in a way that is scary but in a way that is practical for how she approaches the world. While her school district may not switch to renewable energy by the time she enters high school, your home, no matter where you live, provides many opportunities to pair STEM with sustainability.

STEM Activities You Can Do at Home

Perform a home energy audit. No matter the type or age of your home, chances are you and your children can make it a bit more energy efficient. It's possible to make any home completely energy efficient and net-zero or near net-zero, depending on what you do. But to do so you'll have to assess your thermal enclosure, the amount of insulation you have in the home, how airtight it is, the types of fuel and electricity you use, and the efficiency of those types of fuel and electricity.

Just as a carbon footprint calculator can give you a blueprint for reducing your household carbon emissions, a home energy audit can provide you with a plan for making your home more sustainable and thereby reducing your carbon footprint, not to mention your utility bills. While a professional energy audit (often available for free or at a low cost through your utility company) is the most accurate way to do an energy audit, you can do some basics with your children, depending on their age.

Dig out your energy bills and go through several months. The appliances and electronics you use and how you use them affect your energy use and costs.

Consider your lighting. What type of light bulbs do you have, and can you replace them with more efficient choices, such as Energy Star light-emitting diode (LED) bulbs or energy-saving incandescent bulbs?

Look for and list any air leaks by checking for gaps around baseboards and at the edges of walls and ceilings.

You can also check your insulation levels in easily accessible places such as an attic or an unfinished basement. Once you have a list of where your home is losing energy, make a plan based on the most significant energy losses and other factors that will vary depending on your family, including the ability for an energy-efficiency investment to pay for itself and whether you need to hire a contractor to fix something.

Install solar birdbaths. An easy, pretty inexpensive (often less than $20) way to get young children to think about solar power is to add a solar fountain to a birdbath. It amazes toddlers and allows you to start teaching them about energy, and eventually clean energy, from a young age.

Garden. Again, gardening is a simple, easy way to practice STEM skills whether or not you have a backyard. Lots of things grow well on windowsills. Start by thinking with your children about what you want to grow. Look inside your fridge. An edible garden should be full of things you actually eat. After you have a list of possibilities, depending on your child's age, you can have them do research to determine what you can actually grow in your area and if it can grow outdoors or indoors.

Young children can learn the gardening process—by planting seeds, getting their hands in the soil, watering the plants, and watching them grow. Older children can do everything from helping to build a raised garden bed to experimenting with the best things to grow and monitoring and adjusting as needed.

Build a rain barrel (or buy one). A rain barrel is another easy addition for those who live in a house with gutters. The collected water can be repurposed in the garden. If you and your children are handy, you can build one using YouTube tutorials, but it's also possible to buy one for less than $100. You and your child can work together to calculate how much water you usually use in your garden and then how much water having the rain barrel reduces and the cost savings.

Build a windmill. Now that you have the garden and the rain barrel, is there a way to make the collected water flow efficiently to the garden? Many hardware stores sell the parts or even kits to build windmills that you could use to power a pump that connects to the rain barrel.

Create a pollinator habitat. If you're gardening, you could create a pollinator garden full of native plants that will attract and feed bees. More research for your children! But some questions to ask are: What flowers or crops do they notice bees on? And do they think the bees benefit the garden? Adding a bat house or a beehive is another option. Again, there are DIY instructions readily available online for creating both, and your children can help with that and figure out the best placement. In terms of beehives, there are also residential companies that will install and manage beehives on your property, and you and your children can help as little or as much as you want.

Don't worry about getting anything right the first time. We've had an at-home garden for several years, and in the first year, the three trees we planted were the only things we kept alive. We installed a bat house and have yet to have any bats call it home. And I'm still working on divesting our home heating system of fossil fuels. It's a journey, but every time I make an improvement that successfully reduces our carbon footprint or see bees in the garden, I feel more joy and less anxiety about the world I'm leaving. The projects, no matter how small, are visual proof of my knowledge that my child and I can make a difference in our little corner of the world.

10
Community

The little shop sits on the edge of the commercial district in downtown Mystic, Connecticut; it's far enough away from the boutiques and restaurants, and the drawbridge that draws crowds during the tourist season, yet it's close enough to be convenient to pop into when walking around the area. It's here that we stock up on olive oil, nutritional yeast, poppy seeds, and other bulk goods, and where I occasionally settle in at the coffee bar with my laptop for a morning of writing.

Part shop, part café, and part gathering place for the local environmental community, The Ditty Bag is the brainchild of Jason Hine, whom you met in chapter 8. In another life, Hine worked in education and community organizing, but he shifted his professional life after reading an article in *National Geographic* about coral reefs. He started researching the climate crisis. His fears about the world that his two children would inherit spiraled. Before long, he was volunteering with a nonprofit that works with legislators on climate solutions and spending his free time thinking about how he could make a difference. Eventually, he decided to make it the focus of his work.

Hine's zero-waste store sells items like those bulk food goods but also laundry detergent sheets and other household cleaning items, dental floss that you can compost and other personal care items, and even some children's toys, all of them designed to help people rid themselves of packaging waste. But the real goal of the store—its magic, if you will—is the community it creates. The café side of the store is used for organized networking events with local officials or letter-writing campaigns around specific legislation, but often, it's simply a place where people meet by chance and have conversations about how they create a healthier planet.

"What's nice about the store—and I knew this would happen—is that all the progressives, all the environmentalists, all the activists come here, and we network, we have conversations, we plan actions," Hine tells me. "There's different kinds of organizing, but having a home base, a brick-and-mortar, makes all the difference in the world. We can do anything from postcard writing to encourage voter turnout to starting a cleanup. Having a brick-and-mortar allows me to do all kinds of stuff, including bringing people together, that I can't do by myself, and I don't think I would've had the time to do it at home after work."

We are stronger together. Communities—schools, neighborhoods, activist and religious groups, or those that organize online or in a particular geographic region—are vital to mitigating and adapting to the climate crisis because they can effect change and provide support in good times and bad.

"Think of your body as an individual, but then it also has parts: my hand, my eyes, my feet," says Yatibey Evans of Alaska, explaining how she talks about the importance of community with her four kids. "All these pieces are a big part of the whole, and without one of them, the whole has to compensate for the lack of something else. That's what our communities are. If somebody isn't doing their part, then the rest of the community has to make up for that. They have to do more. Each of our unique skill sets, upbringings, and those differences—when we put them all together, I think our communities are able to flourish even more."

I don't need, for instance, to teach my daughter how to can her own produce, but if she knows how to look up from her smartphone and engage in a conversation (even with someone who doesn't always share her viewpoints), she'll be able to learn from someone who knows how to do it, should the need arise. Or—I hope—she'll have a community to help her process traumatic climate events and understand how to talk to and work with her neighbors so there will be fewer traumatic climate events for her to cope with in the first place. Solving the climate crisis will require our children and us to create more collectively minded communities than we currently have.

"Lots of times when we're having big feelings, our instinct is to look away—to numb them, to think, This is bigger than we can handle," says Michele Bigley, the regenerative travel specialist and climate activist. "But we need to remember that whenever anything is hard, we tend to find other people to help hold us up and go through the pain together. It's a way we process."

Our collective resilience will grow if our children can adequately foster a sense of community, banding together on like-minded matters and pursuits. By doing this, we have the chance to improve our fractured planet.

Climate change is a massive collective-action problem. It's not enough to simply reduce our own carbon footprint—system-level change is essential. To mitigate and adapt to the climate crisis, emissions need to be reduced on a massive scale, and that needs to happen through government policy—policy that will be implemented only if there is a critical demand for action from a community of engaged citizens.

"To connect to other people around them," says Rowan Ryrie, mom of two and cofounder of Parents for Future UK, when we chat about the skills she thinks children growing up in the climate crisis need. "To have that kind of strong sense of community and connection to place, and to build those relationships in a way that's really genuine, feels really important."

Yet societies feel more divided than ever in recent history. Surveys of people in seventeen countries with advanced economies in 2020 and 2021 showed that more than 60 percent believed their country to be more socially divided than ever. How do we build collectively minded communities? What skills do children need to create community, and how can parents help develop them?

"I'm not an expert on this and I'm navigating this on a personal journey rather than having any expertise in education or child psychology, but I think it's about resilience—emotional resilience—rather than a specific academic skill set," says Ryrie. "Soft skills. Schools do teach some of that, but it's not necessarily the central focus of the educational system."

While Ryrie is not an expert in child psychology, her background in environmental and human rights law makes her an expert in climate policy, and many of the experts I've spoken to over the years, from Dr. Carla Manly to Sara Peach, editor of *Yale Climate Connections*, to Dr. Holly Parker, have touched on the development of social skills as one of the best ways our children can adapt to the future and the inevitable environmental issues they'll face.

Social skills, those interpersonal abilities, such as the ability to express emotions, communicate, listen, share, cooperate, and resolve problems, are keys to community. They can help children grow into adults who have stronger friendships and better career and educational outcomes.

Social skills encompass a wide range of abilities, such as these:

Communication: Effective communication involves not only verbal but also nonverbal expression: active listening, speaking clearly, using appropriate body language, and maintaining eye contact.

Empathy: The ability to connect with others on an emotional level lets us understand and, ideally, even depoliticize issues like climate change and its solutions.

Conflict resolution: Empathy leads right into conflict resolution, something we need if we want to change individual but especially community and systems behaviors to reduce emissions.

Assertiveness: Expressing opinions, thoughts, and feelings respectfully can help inspire action and help people understand why specific steps aren't being taken.

Emotional regulation: The ability to manage and control one's emotions is crucial, as is the ability to express one's feelings.

Social skills are a bridge between scientific knowledge and real-world action when it comes to climate mitigation. They enable effective communication, collaboration, and advocacy, all essential to addressing climate change's complex and urgent challenge. With strong social skills, it can be easier to mobilize the collective efforts needed to mitigate the impacts of climate change effectively.

"You have to have patience to say, 'Let's talk a little bit more,'" says Parker. "'Let me tell you a little bit more about this story. Let me tell you a little bit more about this. And how can I help you understand the world around you?' That takes knowing your audience and having patience with your audience. Sometimes it's a three-year-old, and sometimes it's a seventy-year-old lobsterman who doesn't want to hear about it."

While Parker makes it clear she's not speaking as a parent, children have played and continue to play a significant role in her life. She's an aunt, was a high school teacher, and is now a college professor. In her work at Bowdoin College, Parker runs the Schiller Coastal Studies Center in the rural coastal community of Harpswell, Maine. The center is trying to create a space where challenging and effective conversations can happen.

"That includes storytelling and story listening," Parker says. "Those are two strategies that I'm really trying to help expose my students to, expose my community to. It's no question that we are in a politically divided moment, and that we need to find a space to have these conversations."

Think tipping point and climate change, and you're probably thinking of degrees like 1.5°C; the temperature the IPCC has long warned us that the planet should not cross. Yet the social tipping point refers to a change in collective behavior. Do you compost your food waste? Don't stress if you don't, but it's estimated that about 28 percent of Americans do. That may not sound like much, but it does mean we've passed the social tipping point of composting. Social scientist Dr. Damon Centola at the University of Pennsylvania has studied how social change happens and believes that once society reaches a threshold where 25 percent of us embrace an idea, large-scale social change occurs. We've seen social tipping points occur regarding marriage equality and antismoking campaigns and even in terms of the renewable energy transition. Can we apply it to climate-change action? Maybe.

While many people are alarmed about climate change, only 7 percent of people in the US say they have joined a campaign to persuade elected officials to take action to reduce global warming. However, 25 percent say they definitely or probably would. That's a sign of great potential to create an engaged climate community that will push for action, creating a social tipping point on climate action.

"If you can get to someone's lived experience, you can find common ground there," says Parker. "We all have moments when we feel anxious. We've all had moments when we felt really excited about the future for our families. We've all had those moments."

Parker may focus on social skills with her students and how they can build community, but as Ryrie mentioned, they aren't something we traditionally focus on.

"I think there's this really interesting decision point in the journey of parenting when suddenly you raise your head from the trenches of the early years of parenting, which are very immediate and the focus during that time is very much the next meal or the next bedtime or the next trip to the park," Ryrie says. "You're then forced to raise your horizons and look ahead in a different way and look at what your kids need to learn in order to equip them, to help them be resilient and happy people for the future they're coming to."

Despite her background in environmental law, Ryrie worked on climate issues only once she started thinking about making long-term decisions for her children's future, starting with where to send them to school.

"That to me is a really inherent quality of parenting, and that really became apparent when I started to make decisions about my elder daughter's schooling," she says. "Suddenly you have to be able to look ahead and make decisions about what your children need to equip them for the future that they're going to live in. How do I make decisions for five years or ten years, and then put them beyond that eleven-year time frame for school?"

Seeing a complete absence of long-term decision-making within society, particularly around environmental issues, spurred Ryrie to act.

"I was interested in how we join some of those dots and how we help parents understand that there is this need for the kind of decision-making that parents do every day to be brought more into the political space, and to have more of a conversation around long-term decision-making on environmental issues, because it's going to impact all of our children's futures," Ryrie says.

She attended some climate marches, read the IPCC report, and decided to get involved. Committed to making the movement more intergenerational and finding ways to bring adults on that journey alongside youth activists, she helped set up Parents for Future UK and Parents for Future Global.

"It became clear to me that it wasn't anywhere near good enough to wait for our kids to take this issue on," Ryrie says. "There's a really important element that we as parents need to be stepping up and taking action on now, because unless we as a generation have some agency and power in the world—political power, economic power, et cetera—and do something now, our kids will face a lot more harm down the line."

Over the past several years, Ryrie has created a community of parent climate activists who are pushing for changes, while also focusing on helping her own child create community.

"It's about helping them connect with the world around them," says Ryrie. "Having that relationship with nature feels fundamentally important, and to be able to connect, to listen to themselves, not just to what the world around them is telling them to value, and to connect to other people around them."

For young children, something as simple as one-on-one time with someone helps them practice social skills. As they grow, everyday things they're already doing, like going on playdates and playing team sports, are also ways to practice social skills. We can also teach our children to recognize and express their emotions, encourage them to talk about their feelings, and listen when they share. Encouraging children to share stories and experiences with family members or friends helps them develop their storytelling skills and fosters meaningful connections.

Remember, resilience in a community does not always mean actionable change; sometimes it means habituating.

The most recent IPCC reports have focused on adaptation as much as on mitigation. Yet climate adaptation will be possible only if we have resilient communities.

Social skills are the glue that holds communities together. They enable effective communication, foster trust and cooperation, and promote a sense of belonging and shared responsibility. Communities that prioritize the development and practice of social skills tend to be more resilient, inclusive, and successful in achieving their goals and objectives.

That's the thing, isn't it? The feeling that others know us, that they're on this journey, too, that even though the paths aren't ever exactly the same, we belong. We can't hide the climate crisis from our kids, and we really shouldn't try to. We're teaching them how to live, and that includes facing scary things. But by creating space for their feelings to matter, for their imaginations to run wild, and for places to return home to, we're creating the communities that will help them face scary things.

And that's what parents want—for their children to thrive no matter what the world around them looks like.

■ 1 ■ Activism

The churros arrived warm and drizzled with caramel sauce. They were a bribe. Or, as one of my daughter's favorite teachers would say, an external motivation. For what? I wasn't really sure; I just knew I wanted to attend the brainstorming climate legislation meeting and I didn't have childcare for the evening. A state politician thought she could get something passed in the next session, and I was hoping that churros would buy me a few minutes of my child sitting quietly at the taqueria so I could participate. I also wanted her to have a glimpse of how the process works, and I wanted her presence to remind us of the future we're fighting for.

It's exhausting and infuriating that it has come to this—that everything from showing up to a meeting to how we brew our morning coffee to how we get to the grocery store is a decision that contributes to what we can save on our planet. The climate crisis forces us to ask ourselves so many questions: Should I give up meat? Should I buy only organic? Should I buy only local? Should I put solar panels on my house? Should I try to get my child's school to compost? What about shampoo bottles? They're made of plastic; should I use shampoo bars?

The list of individual actions we can take can feel endless, and attempting to figure out which solutions to put our energy toward can be so overwhelming that we are frozen into inaction. Despair, after all, is easy; hope is hard. But as parents, we need to know that every one of us can do *something*. Small steps do matter and can lead to hope and resilience.

How can we help show our children (and ourselves) that our actions matter?

Through activism. It's a word that gets thrown around a lot these days. Yet activism helps children practice social skills in real life, including how to work on a team, and how to communicate and plan with others effectively. It also gives people a sense of purpose, a community, and a supportive network, helps them create healthy habits, and helps them adapt to setbacks. While not all children can be Greta Thunberg (nor should they be), activism can be a powerful tool for building resilience. And vice versa: Resilience can be a powerful tool for becoming an activist.

"We have to do something to work to change the problem," Michele Bigley, the regenerative travel journalist and climate activist, tells me when we chat about her activism. "Our whole life as parents is offering our kids the skills to be the people they have the potential to be, but also pulling back and modeling the adult we want them to be. You could tell them and tell them and tell them, but as young people they most likely are going to mimic what we do and take on some of those values because it's a part of how we live."

Some children and adults will naturally gravitate toward activism; some will not. As an introvert, I tend to shy away from activism. It took prodding to get me to attend those brainstorming sessions. But even a tiny amount of activism can increase a person's well-being, and among children, youth involvement in activism is associated with fewer health problems. Studies have shown that activism can benefit physical and mental health, reduce depression, and lower the risk of substance abuse. Civic-minded teenagers have also been found to achieve higher income levels as adults than non-civic-minded teenagers.

"I love giving back to the environment, and when I see all the trash before doing a cleanup and then see it after, I see how much of a difference we can make," Cash Daniels, the thirteen-year-old from Tennessee, tells me.

Cash, you may remember, began regularly doing beach cleanups and organized students worldwide to pick up a million pounds of litter before the end of 2022 after realizing litter on a beach could harm animals.

"I do consider myself to be an environmental activist," Cash says. "I fight for the environment, and I raise awareness for all—for animals that can't raise a voice for themselves."

What does it mean to be an activist at age three? At thirteen? At thirty-three? And what does it look like to let your child be one?

"Unfortunately, it's not enough for us just to have the practices that we have as individuals and as a family. More has to be done," says mom of three Alexina Cather.

Activism looks different to different people. For some, it's attending rallies and organizing awareness campaigns. For others, like Cash or seventeen-year-old Clare Flaherty or some of the people you've met throughout this book, it's having an idea of a solution to one aspect of the climate crisis and working to implement it. In the cases of

Science Moms and Parents for Future, it's pushing legislation, talking with legislators about climate-change mitigation, and gathering support from others. An activist is a person who uses or advocates for actions in support of or in opposition to an issue, aiming to bring about foundational change.

"I've taken my kids to protests for everything from the climate to immigration to social justice," Cather says. "They've gone to marches, and they understand why it's important to use their voice, but it's a balance of how to talk to people, get your message across, and inspire people to action. We know the climate crisis is overwhelming, and people tend to just shut down a lot around climate issues, but we're moving into the stage, especially with our oldest son, of inspiring them to be a little bit more active and impactful in their community."

Climate activism encompasses a wide range of actions and strategies aimed at mitigating and adapting to the climate crisis. Likely, some of the first actions that come to mind, especially if you have kids, are climate strikes and marches.

Greta Tintin Eleonora Ernman Thunberg is a household name these days, but that wasn't always the case. Greta was in elementary school when she first learned about climate change, and, wanting to do something, she became a vegan and began refusing to travel by plane. In August 2018, before the Swedish elections, she missed school to sit outside parliament with a sign reading "Skolstrejk för Klimatet" (School Strike for Climate). It was a summer of heat waves and forest fires and the first time climate change was seen as a key issue in a European election. Over several weeks, she was joined by more and more people, and while after the elections she went back to school, she began skipping classes on Fridays. Eventually she founded Fridays for Future, a youth-led and youth-organized global climate-strike movement that aims to make political leaders take action to reduce greenhouse gas emissions.

Climate strikes, protests, and demonstrations are just some of the various types of activism that kids and parents are undertaking. They aren't suitable for every child; some, like Clare, focus on environmental education.

"When I was in elementary school, my sister got Lyme disease twice and then my brother got it," says Clare. "I found out tick-borne diseases are exploding as a direct result of climate change. That piqued my interest, but as I started thinking and learning more about climate change, I understood that there are changes we could make in our own home and that would make a difference in our community."

Clare started running educational programs in her Rhode Island community, such as No Child Left Inside, which incorporated climate-change trivia while trail hiking and picking up trash.

"Many of the programs I initially ran had a simple formula where you're having fun and being in nature but also educating," says Clare. "Now my climate activism has morphed to bigger presentations and programs, but they all stay true to the core of education." Clare now works with the Climate Initiative, an organization working to develop a youth voice that influences decision-makers to embrace climate solutions.

It's a similar process to what Science Mom Erica Smithwick took. She's been on the faculty at Penn State for more than a dozen years researching how wildfires affect ecosystems, and she also looks more broadly at questions around carbon storage in natural systems and the role of forests and other natural solutions in mitigating climate change in Yellowstone National Park, Ghana, and other places.

"Increasingly, I've been interested in working with people, because people are the core of eventual solutions," Smithwick says. A few years ago, she joined Science Moms to educate more people about her work and solutions.

"I'm a mother of three children, and I was excited about the opportunity to become involved in Science Moms because it really is who I am," she says. "My identity as a mother is my identity as a scientist. They're interchangeable. When I'm thinking about my fieldwork and thinking about the role of science and society, I'm also thinking about the future that my children will inherit. I know that sounds a little stereotypical, but it's really true. I study the same landscapes that my children will eventually inherit. I think it's important that I do it right."

For many, activism often starts within their own communities. Remember those school districts using renewable energy? Some of them switched because parents and students like Nha Khuc asked them to.

"We were having a career panel as part of a school project and my task was to interview a woman, Mikhaila Gonzales from Spark Northwest, which helps communities get solar, and I asked, 'What would it look like for a public school to get solar panels installed?'" Khuc, who is now a student at the University of Washington, tells me.

Khuc turned that initial question into getting her high school to install solar panels on the roof. Over the course of two years, she worked with about nine other students, reached out to other schools that had switched to solar energy, created a presentation for her school board, lobbied them, and helped raise money for the solar panels.

"I had never done a project of this size before," Khuc says. "I'd never done a community project before, and I never thought I had the ability to do this. I'm from Vietnam, and I'm self-conscious about putting myself out there and doing things, but I was surprised by the community support. I felt a sense of community, and it really kept me motivated."

Here's a brief overview of the steps you and/or your child can take to make a difference:

1. **Gather your squad.** Invite children and/or other parents over for a brainstorming session. Talk about why you want the schools to switch to clean energy, outline the benefits, discuss what you see as challenges, and see who is interested in helping.

2. **Research.** You're going to need to understand where the school district currently gets its energy, how much energy it uses, and how renewable energy works. You should be able to ask the school administration for its figures on energy use, or they're likely available in a school board budget. Figure out your school's options for renewables—solar, wind, or something else—how much it will cost, and how much energy the school district could reasonably expect to produce. It's also wise to look for examples of other schools already using renewables that you can cite. At this point, depending on where you live,

you can reach out to organizations such as the Sierra Club or Generation180 in the US or Solarsense in the UK, which can often assist with research and give suggestions to help develop a renewable energy plan.

3. **Define your goal.** The goal should be for your school to get 100 percent of its energy from renewable sources, but you may need to set a timeline for that as well as percentages that are coming through renewables directly on school property versus from public and/or private electrical systems, depending on the setup where you live.

4. **Get community support.** Next, it's time to amass support. Students should not only ask their fellow students to support going solar by signing a petition but also approach groups such as the PTA, nonprofits focused on youth, religious institutions, and any other local businesses and organizations that they believe would be interested. The more you can show community buy-in, the more likely it is that the plan will work, especially in public schools, where taxpayers ultimately fund the schools' energy system.

5. **Set up a meeting.** Once you have your information gathered, ask to speak to the school board. You could make an appointment with the head of the school system or board, but ultimately you want to ask for a time to present your plan at a meeting. Inform your supportive community members of the school board meeting and ask them to show up for it.

In the best-case scenario, the school board will be entirely ready to move forward. More likely, however, it will have follow-up questions, more research will need to be done, and a vote by the local town might even need to take place, depending on the structure of the local government.

"I was surprised how long it takes to do a project," Khuc says. "Our timeline was for about a year, and it ended up taking two. There was so much back-and-forth between different people. It gave me a new perspective."

Encourage your children to continue to follow up. It's an excellent example of working toward a larger goal by practicing skills like creative problem-solving, building community, and persistence. Finally, if the school cannot move toward renewable energy right now, allow your kid to feel defeated. They and you need to process that loss, but then ask, "So what can we do now?"

Let your child come up with alternative solutions. Maybe it's going back to the drawing board and proposing a new mix of renewables, perhaps it's looking for funding sources if cost is an issue, or maybe it's even supporting new school board candidates who are more in line with moving toward renewable energy. This isn't something you and your children can fail at unless you give up.

"You have to keep an open mind," Khuc says. "We did a lot of outreach where we didn't get a reply, so we had to think about 'How can we move forward?' so we wouldn't get stuck." Khuc adds that for her, one of the best parts of the project was that she became more confident in herself, and more willing to speak her mind and work through challenges.

Here are some other ideas for climate action:

Plant something: "My friends have a lot of feelings of helplessness," says Tamar Shoshan, the New York City teenager. "We have the debate: Should individuals be doing small things to help the climate crisis? Or should we rely on corporations? A lot of the time, we don't know where to get started, or [we think,] *If we do anything, is that really going to have an impact?* Something that I really like is encouraging people who feel helpless to plant something—anything—because just having that connection to nature makes you more open to caring about it."

Environmental education: Like Clare, you can help raise awareness about environmental issues through workshops, educational hikes, and other educational campaigns. Local libraries are often looking for people to hold talks and workshops on different topics. Think about something you're passionate about. Maybe it's single-use plastics, for instance. Could you put together a presentation on the issues and solutions to some common single-use plastics?

Online activism: Science Moms uses its online and social media platforms to educate its followers about climate solutions and how to talk to others about climate actions.

Talk: We keep coming back to it, but talking about the climate crisis is one of the best actions individuals can take.

Consumerism: You can always be an activist with your money, by supporting companies that engage in sound environmental practices. And speaking of money, you can start with your bank. Banks use deposits to make investments in all kinds of projects. Your bank deposit could fund investments in local agriculture, solar farms—or in oil or gas. According to nonprofit community bank cofounder Kat Taylor and environmentalist Bill McKibben, when you bank with one of the top four banks (Chase, Bank of America, Citigroup, and Wells Fargo), having $125,000 in a savings account is equivalent to an entire year's worth of the average American's carbon impact. That's because we often think that when we deposit money, it just sits in the bank, but the reality is that banks use that money to make money by lending it to other places like fossil fuel companies. These banks are some of the biggest investors in fossil fuel projects. There are online tools that help consumers determine whether banks, mutual funds, and other financial products invest in fossil fuels.

Lobbying and advocacy: I try to email my local legislators about once a month on an issue related to the climate. I want them to know that I care and that I'm paying attention to how they vote.

Workplace: Remember the steps the students took to get solar panels at their schools? You can do something similar within your own office. First, consider what you're good at. Maybe you have a clutter-free, organized desk, or you're a coffee enthusiast, or you're particularly good at saving money. Now think about how you could use your passions and skills to do something at work that helps mitigate the climate crisis. Can your company reduce its plastic waste from Keurig? Can solar panels be added to the building? Can a travel policy that considers the impact of travel on the planet be adopted? There are a million tiny ways you can work on making your office a little less climate-intensive, and you can talk to your kids about your actions. It's essential that they see you taking action.

"I cannot understate the [importance] of taking action," says clinical psychologist Dr. Carla Manly. "We are showing children by our actions that we care about their future. I care about the planet. I'm not just going to trash it while I'm here and leave it to you." Even if and especially if at first your action fails, you're modeling resilience, problem-solving, and creativity for them.

Community-based conservation: Participate in local conservation projects, such as cleaning rivers, beaches, or parks and preserving natural habitats. Cash suggests using his organization, the Cleanup Kids, as a starting point. "Sign up to be a cleanup kid, and then you can go out, take pictures of all the trash that you see, and log it. And we'll know how many pieces you've cleaned up," Cash explains. It may seem like a small thing, but there's evidence that beach cleanups rapidly and dramatically reduce the amount of microplastics in the environment. Scientists at Norwegian research organization NORCE found that within a year of removing bags, bottles, and other large plastic litter from the shore of an island near Bergen, there was a 99.5 percent reduction in microplastics on land and in surrounding oceans.

To be effective, activism requires curiosity about the world, understanding a problem and a possible solution, effectively engaging others with the problem and solution, and the ability to not only bounce back from setbacks along the way but reevaluate your strategy, adjust, and try again.

"I certainly don't want to force my child to go out there, but I think they understand the value of advocacy," says Cather. "At some point we're going to have the conversation, which we'll have as delicately as possible, about how this is the world they're growing up in. It's not at all going to be the same as when they're our age. And it's heartbreaking as a parent to know that the choices of our generation, and those of generations past, are going to take things away from the future generations. They need to know that it is up to them, that they're going to have to fight even more than we are."

Greta and other youth activists reinvigorated attention around the climate crisis. Still, hearing her story, I can't help but think of her parents and wonder at the scaffolding and level of trust they built in their relationship with her.

"As parents we have to get comfortable with discomfort," says Dr. Tamara Yakaboski, the educator and climate-crisis activist, who thinks about the way she has slowly let her kids have more freedom with sleepovers and bicycling. "I think a lot about scaffolding and building trust and if we've given them enough skills to know their capacity, to know the limits, to know when to bail, to know when to call for help and who that help is, which looks very different based on identities and city climate and whatnot."

The summer her daughter turned ten, bicycling was a big deal. "She wanted to ride her bike in the neighborhood," Yakaboski says. "Here in Colorado, we're very bike friendly, but we'd had a record number of bicycle-car deaths that year. I was and still am really grappling as a parent with 'I want you to have independence and freedom and take some risks within boundaries, but I don't trust the adults in the world.' I ended up letting her bike to one of her summer camps, which was a mile away, and pretty straightforward. She knew the path."

Yakaboski created different points along the way and would bicycle with her daughter to one end, slowly widening the difference from bicycling behind her along the whole path to being a block and a half away.

"I wanted to watch her bike choices, and then we talked about safety and how it went. Then I would slowly be like, 'Okay, I'll bicycle to this point and let you carry on.' Kids have to feel that we trust them to make good choices, trust them to know what to do."

You blink, and the child you stressed about buckling into the car seat the right way is five and going to school for the first time. Blink again, and they're a ten-year-old wanting to bicycle to summer camp by themselves, and then all of a sudden, they're fifteen and wanting to protest outside an embassy.

"When I was having this internal debate about how I help [my daughter] get independence, I did a lot of research, and one of the things that stood out to me was research in the *Journal of Pediatrics* by Peter Gray," says Yakaboski. "He talks about research on the play deficit and how it's correlated to a mental-health decline in children, the loss of freedom and the loss of risk-taking, and that instead of creating a bubble for our kids, we need to create boundaries and containers."

Gray and others have argued that just as the risk has been taken out of many playgrounds, and as children's free play and play with other children have declined sharply over the past half century, anxiety, depression, suicide, feelings of helplessness, and narcissism have increased sharply in children, adolescents, and young adults. Free play, according to Gray and other researchers, helps children learn how to make decisions, solve problems, regulate their emotions, get along with others, and even experience joy.

And free play is a building block for allowing kids the freedom to be activists.

"My kiddo has been at marches with me since she was little, so it's familiar and we live in a relatively safe town," says Yakaboski. "Our marches are pretty mild, but it's hard for me to perceive how I'll feel as she gets older, just knowing how this biking around the neighborhood thing feels so conflicting."

As a four-year-old, my child was already craving the type of freedom Yakaboski's ten-year-old wanted, asking when she could have a sleepover and walk into town by herself. At the same time, she's the kid who will explain why your gas car isn't good for the planet or that bees aren't scary but helpful. I can imagine in a few years she'll want to participate in a march, organize a beach cleanup, or plan an educational hike like Clare. Like Yakaboski, I'm unsure how to handle that, but I asked Manly her thoughts.

"I think it's important for parents to look at what is truly unsafe and the risk for the child, rather than letting their fear get in the way," says Manly. "If I'm saying no to Sarah about going out and doing a march downtown and it's a peaceful march, I might be uncomfortable with that, but she will be fine. The chances of her getting hurt in a peaceful march—of her getting hurt in a beach cleanup, especially depending upon her age and if it's well supervised—are small."

Manly suggests returning to those curiosity and problem-solving skills to say, "Hey, I'm not sure I'm comfortable with this. Do you mind if I'm the one who drops you off? Do you mind if I go too? I can hang out in a quiet place by the statue if you need me."

"Sometimes parents say no because it's simply not in their wheelhouse or it's not something they're interested in or comfortable with," Manly says, "but if you do say no—and there will be times you will say no—always offer an alternative."

Most of us love choice. Even if you can't let your child go to San Francisco or New York City to do this big protest, you could say, "I'll help you organize this one in our downtown. I'm behind that." In other words, helping to support the child's need and desire to do something, and honoring that, supports the child's sense of agency, whether it's attending a march, having a sleepover, or bicycling to summer camp.

"Parents also need to realize there are times when you will have to say no," says Manly. "If somebody wants to go and set fire to a bridge as a protest, well, that'll be a hard no."

You've probably seen the news about climate activists doing everything from disrupting tennis matches to throwing soup at Vincent van Gogh's *Sunflowers* at the National Gallery in London to smearing paint on the case and base of an Edgar Degas sculpture in the National Gallery of Art in Washington, DC. I have a lot of mixed feelings about those tactics. How would I feel if my child did that? I'd understand why they did it. I'd understand at least some of the emotions behind the actions. These days, my hometown has this party-at-the-end-of-the-world vibe. It's become a sought-after tourist destination, and some days, watching the tourists wander around, the increased traffic, litter, and prices they've created, while knowing that all that tourism creates more carbon emissions, makes me want to scream. So I get it. Yet there's evidence that although direct action and civil disobedience, even in nonviolent forms such as sit-ins or blockades, draw attention to environmental issues, they might undercut public support for the movement. And there are, of course, personal costs to them. Tennessee-based soil scientist Rose Abramoff has been arrested for her climate activism, which has included chaining herself to a drill on a fossil fuel pipeline. She is believed to be the first climate scientist to risk felony charges for a climate protest. She was also fired from her job at Oak Ridge National Laboratory in Tennessee for participating in a previous climate protest.

It's easy to judge property destruction and civil disobedience from afar, but I don't think any of us know what we would do when faced with specific situations. In 2021, Swedish political theorist Andreas Malm wrote the book *How to Blow Up a Pipeline*, which was later turned into a film directed by Daniel Goldhaber.

The premise is simple: We know we need to reduce fossil fuel emissions, but we're not moving fast enough, so is it time to consider more extreme actions? Malm's book and the film aren't advocating for violence—a line is drawn there—but they do consider whether property destruction, including of fossil fuel infrastructure, is necessary. One of the scenes from the film that strikes me is when a group of young people sit around talking about whether they're terrorists.

"Of course . . . ," one person says. "We're blowing up a goddamn pipeline!"

"They're going to call us revolutionaries," says another. "Game changers."

"Not so," another says. "They're going to call us terrorists. Because we're doing terrorism."

(If you haven't read the book or watched the film, I encourage you to do so.)

How social movements succeed or fail in pushing their agenda in society is a highly complex dynamic that we don't fully understand. Remember, there's research that says that once a social movement gets to 25 percent, that's the tipping point for mass action, yet how we perceive protest largely stems from a shared sense of right and wrong.

"You would think that people would show a great deal of deference to activist groups that are representing their political or demographic group and not be negatively influenced by their extreme tactics, but that's not what we found," Robb Willer, a professor of organizational behavior at Stanford Graduate School of Business, writes. Willer is one of the authors of a paper about research on the widespread claim that extreme protest actions—those viewed as "harmful to others, highly disruptive, or both"—erode public support for movements.

For all the good activism can do, any type of activism—but especially climate activism—can be a tricky thing to balance. It's easy to get burned out, and many climate activists report experiencing unhealthy aspects of occupational identity. That makes sense when you start to see all the ways even the most eco-conscious among us are often forced to live in ways that add to human-made climate emissions. It can be hard to balance living your values with being a part of today's consumer-driven culture and maintaining relationships with family and friends.

"You feel like you're a Debbie Downer all the time," Cather says. "But this time, with our third kid, I had to say to people, 'I don't want you to buy anything for this baby. I will get everything that I need secondhand.' Or 'We still have clothes from the other kids,' or 'We just don't do plastic toys.' But there are arguments and disappointments with family members over that."

.

Many climate activists, including most of the ones I spoke to, also deal with their own stress and grief about having witnessed or experienced climate-related loss. If you're feeling burned out, or your child is, remember to limit your time spent doomscrolling and spend time in nature. Check in and reach out for professional mental health support if you need it.

How Activism Can Help with Climate Grief

Despite its difficulties, climate activism can help with climate grief and anxiety and help people find their community. You can start with gratitude.

In *I Want to Thank You*, Gina Hamadey chronicles a year of writing 365 thank-you notes to everyone from friends to family to former colleagues to casual acquaintances. She found the letters boosted her confidence and connected her to others and herself. Just as talking about the climate crisis is important, so is thanking people for their actions. These days I send people a thank-you note when I notice them taking a climate action, be it to a news anchor who, after reporting the news about the heat, took a moment to comment on how it worried them, adding emotion that's often lacking in news reports on climate change; or to a hotel that had reusable coffee mugs at breakfast. All small things, but all meaningful. Plus, gratitude, numerous studies have shown, allows us to be more creative and optimistic and builds resilience. Numerous studies found that those who performed a gratitude task reported increased emotional coping skills. It helps us to recognize the positive instead of focusing solely on the bad.

"I have fears for the environment, and, like, animals going extinct," says Cash. "That's why I'm doing it: to keep those animals from disappearing." In his own river in Tennessee, Cash has removed almost 18,000 pounds of trash by doing cleanups once a month.

"I think lots of times people, especially kids, think they don't have any power to make any change, so it's good to see that you don't feel that way," he says.

Showing up, being involved in your circle, using your voice to take advantage of opportunities to connect are powerful.

"They see the responses we get from volunteers and others attending events saying that we can do something," says mom Stephanie Rice, who has brought her children to a number of events advocating for climate action. "I feel it's a positive message because we're not just accepting things as they are; we're saying that we can, through our actions, through our voice, through our representation, show up and actually change the way things are." Rice adds that she knows it's not always possible, especially for single parents, to show up to events, but that she's found many different ways to show support, including signing petitions.

Activism helps you share your values with the world and increase your social connection and sense of purpose.

I'm not a joiner. I'm more comfortable reporting on processes than being a part of them, but I know that one of the best ways to create the world I want for my child is for her to see me live it. That means getting involved, failing at times, and bribing her with churros.

Conclusion

The year my child graduated from preschool, a haze of smoke settled over our coastal town in the late spring. With a million thoughts running through my mind that week, I was having difficulty sleeping. Up early to witness the sunrise, watching as the sky turned red, I couldn't help but whisper an old saying: "Red sky in the morning, sailor's warning."

Climate scientists have been issuing dire warnings for decades. But now many of us are experiencing the effects of climate change firsthand. That year, on the day of my daughter's first soccer practice for the summer season, I could smell the smoke from wildfires hundreds of miles away. I couldn't help but think, *There's supposed to be more time . . .* before I have to make my child wear a mask to practice, before I question the coach's decision to keep practice outside, before I wonder about skipping it. There's supposed to be more time before the climate crisis becomes a part of my child's daily life.

Of course, I knew it would eventually come to this—I knew my daughter would have to navigate the climate crisis for much of her life. Already, according to UNICEF's Children's Climate Risk Index, more than half of the world's children live in countries at high risk for climate and environmental hazards. Still, I had hoped, selfishly and foolishly, that the place of her birth would give her more time.

How will I answer if she does one day ask me, "How could you have had me, knowing what you knew?"

The decision to become a parent is an incredibly personal privilege. Many people are considering either having fewer children or not having children at all because of issues related to climate change. Four in ten people aged sixteen to twenty-five worldwide said in 2021 that they were hesitant to have children because of the climate crisis, according to a study of 10,000 people published in *The Lancet*. Even when a child is wanted, little is known about how climate change and heat may impact fertility. Given what we know about the world our children will face, how does one decide to have a child?

The truth is that I never considered not being a parent. I couldn't imagine not being a mom, and not to have become a parent because of the climate crisis would have felt like giving up, an acknowledgment of doom. As climate justice writer Mary Annaïse Heglar has said, "Don't worry if we're doomed or not. Worry about what you stand for and stand for it."

This is our place in history. We didn't get to pick it, but we do get to influence the world in which our children and grandchildren will live.

So what do I stand for? For raising a kind, considerate, curious child who will be a determined and flexible problem-solver, sure of what she stands for but willing to consider others' points of view. I stand for supporting people who are supporting the planet. I stand for leaving the Earth a better place than it was when we received it. For hope, joy, and courage.

"My child makes me hopeful," Dr. Rosimar Rios-Berrios, the Science Mom and scientist at the National

Center for Atmospheric Research in Boulder, Colorado, told me years ago.

That was true for me that late-spring day, and even as I donned a mask for preschool graduation, it remained true.

This does not mean I think everything will be fine. The fires in Maui, the floods in Libya, the loss of life: We're going from one catastrophic event to another, and it's going to happen faster now. Trips to the supermarket still give me anxiety—all that single-use plastic. Watching my daughter watch a bee outside, I am often overwhelmed with grief and anger. There are still many days when I feel there are no good choices and that we're out of time. But then I'll look at my daughter running up and down the beach near our home, delighting in how the water laps at her feet, and chases her if she runs back from it, and I can't help but think that she and her friends are saving the world.

It shouldn't be their job to save it, of course. It shouldn't be ours, either. Yet I believe children are the ultimate manifestation of hope. Having my daughter has let me be a kid again and has renewed and refocused my energy on fighting for the future. Our children, after all, are a bet on a future we can't yet imagine and can't fully control.

Our children need for us to be both optimistic and courageous. They need to know we fought as hard as we could for as long as we could.

"We can't solve everything overnight, but we are in a position to make serious changes and avert the worst of climate change," says Dr. Joellen Russell, the Science Mom, mother of two, and University of Arizona professor. "I'm truly inspired by the everyday parents I talk to and the students in my classes, and have faith that we will take the actions necessary to provide future generations with a healthy and safe planet."

There's a balance to be found between helplessness and cautious optimism. I worry less about the things I can't control. I can't, for instance, make governments act or make corporations factor the environment into their product-making decisions or even stop my neighbors from spraying toxins on their yard. I am still forced to do things I know are harmful to the planet, because of how the system is set up. I live with sadness, anxiety, and anger. Yet I focus more on creating and seeing joy, on what my family can do, what the community I live in can do, and what the community of parents and climate fighters worldwide can do.

We're fighting for the fireflies. Since spotting them outside my daughter's window, we've embarked on efforts to transform our yard into a firefly sanctuary. Many of my childhood reference landscapes, like the shoreline I grew up with in the Outer Banks, feel like things we will lose. I hope that decades from now, if we act, if we are courageous, our children's reference landscapes will have only grown. I hope that when she is an adult, my child will see more fireflies. I hope that years from now, our kids' grandkids won't think about the climate crisis, because we took the climate actions we needed to, and that they will live with an awareness of and care for the environment that are second nature. I hope that they go outside and are awed by fireflies, trees, and the moon in the sky.

Additional Reading

BOOKS

Active Hope: How to Face the Mess We're in Without Going Crazy, by
  Joanna Macy and Chris Johnstone

Braiding Sweetgrass, by Robin Wall Kimmerer

The Climate Book, by Greta Thunberg

Climate Change and Happiness, a podcast series, by Thomas Doherty
  and Panu Pihkala

*A Field Guide to Climate Anxiety: How to Keep Your Cool on a
  Warming Planet*, by Sarah Jaquette Ray

How to Blow Up a Pipeline: Learning to Fight in a World on Fire, by
  Andreas Malm

*I Want to Thank You: How a Year of Gratitude Can Bring Joy and
  Meaning in a Disconnected World*, by Gina Hamadey

*Live in a Home That Pays You Back: A Complete Guide to Net Zero and
  Energy-Efficient Homes*, by Anna DeSimone

*Planet Palm: How Palm Oil Ended Up in Everything—And Endangered
  the World*, by Jocelyn C. Zuckerman

*A Pocket Guide to Sustainable Food Shopping: How to Navigate the
  Grocery Store, Read Labels, and Help Save the Planet*, by Kate
  Bratskeir

The Tantrum That Saved the World, by Megan Herbert and Michael E.
  Mann

*When the World Feels Like a Scary Place: Essential Conversations for
  Anxious Parents and Worried Kids*, by Dr. Abigail Gewirtz

The World Is Ours to Cherish: A Letter to a Child, by Mary Annaïse
  Heglar

ONLINE RESOURCES

Climate Emotions Wheel, from the Climate Mental Health Network
  (climateadvocacylab.org/resource/climate-emotions-wheel)

Climate Kids, from NASA (climatekids.nasa.gov)

50 Things to Do Before You're 11?, from the National Trust
  (nationaltrust.org.uk/visit/50-things)

"How to Turn Climate Anxiety into Action," Ted Talk by Renée
  Lertzman (ted.com/talks/renee_lertzman_how_to_turn_climate
  _anxiety_into_action?language=en)

Local Catch Network (localcatch.org)

Recycle City, from the US Environmental Protection Agency
  (epa.gov/recyclecity/)

Sources

INTRODUCTION

Flavelle, Christopher. "Tiny Town, Big Decision: What Are We Willing to Pay to Fight the Rising Sea?" *New York Times*, March 14, 2021. nytimes.com/2021/03/14/climate/outer-banks -tax-climate-change.html.

Masson-Delmotte, Valerie, Panmao Zhai, Anna Pirani, Sarah L. Connors, Clotilde Pean, Yang Chen, Leah Goldfarb, et al., eds. *Climate Change 2021: The Physical Science Basis: Working Group I Contribution to the Sixth Assessment Report of the Intergovernmental Panel on Climate Change.* Intergovernmental Panel on Climate Change, June 2023. Cambridge, UK: Cambridge University Press. doi.org/10.1017/9781009157896.

1: TALK

Climate Investigations Center. Climate Files: Exxon Documents. climateinvestigations.org/exxonknew/.

Franta, Benjamin. "What Big Oil Knew About Climate Change, in Its Own Words." *Conversation*, October 28, 2021. theconversation .com/what-big-oil-knew-about-climate-change-in-its-own-words -170642.

Goldberg, Matthew H., Sander van der Linden, Edward Maibach, and Anthony Leiserowitz. "Discussing Global Warming Leads to Greater Acceptance of Climate Science." *Proceedings of the National Academy of Sciences*, July 8, 2019. doi.org/10.1073/pnas .1906589116.

Goldenberg, Suzanne. "ExxonMobil Tried to Censor Climate Scientists to Congress During Bush Era." *Guardian*, May 25, 2016. theguardian.com/business/2016/may/25/exxonmobil-climate -change-scientists-congress-george-w-bush.

Hall, Shannon. "Exxon Knew About Climate Change Almost 40 Years Ago." *Scientific American*, October 26, 2015. scientificamerican.com/article/exxon-knew-about-climate -change-almost-40-years-ago/.

Kurland, Zoe, Katie Hafner, Elah Feder, and the Lost Women of Science Initiative. "The Woman Who Demonstrated the Greenhouse Effect." *Scientific American*, November 9, 2023. scientificamerican.com/article/the-woman-who-demonstrated-the -greenhouse-effect/.

Leiserowitz, Anthony, Edward Maibach, Seth Rosenthal, John Kotcher, Parrish Bergquist, Matthew Ballew, Matthew Goldberg, et al. *Climate Change in the American Mind: April 2019.* Yale University and George Mason University. New Haven, CT: Yale Program on Climate Change Communication, June 27, 2019. doi.org/10.17605/OSF.IO /CJ2NS.

Levy, Adam. "Scientists Warned About Climate Change in 1965. Nothing Was Done." *Knowable Magazine*, May 30, 2023. knowablemagazine.org/content/article/food-environment/2023 /scientists-warned-climate-change-1965-podcast.

Masson-Delmotte, Valerie, Panmao Zhai, Anna Pirani, Sarah L. Connors, Clotilde Pean, Yang Chen, Leah Goldfarb, et al., eds. *Climate Change 2021: The Physical Science Basis: Working Group I Contribution to the Sixth Assessment Report of the Intergovernmental Panel on Climate Change.* Intergovernmental Panel on Climate Change, June 2023. Cambridge, UK: Cambridge University Press. doi.org/10.1017/9781009157896.

Milman, Oliver. "Revealed: Exxon Made 'Breathtakingly' Accurate Climate Predictions in 1970s and 80s." *Guardian*, January 12, 2023. theguardian.com/business/2023/jan/12/exxon-climate-change-global -warming-research.

Nuccitelli, Dana. "Scientists Warned the US President About Global Warming 50 Years Ago Today." *Guardian*, November 5, 2015. theguardian.com/environment/climate-consensus-97-per-cent /2015/nov/05/scientists-warned-the-president-about-global-warming -50-years-ago-today.

Pörtner, Hans-Otto, Debra C. Roberts, Melinda M. B. Tignor, Elvira Poloczanska, Katja Mintenbeck, Andrés Alegría, Marlies Craig, et al., eds. *Climate Change 2022: Impacts, Adaptation and Vulnerability: Working Group II Contribution to the Sixth Assessment Report of the Intergovernmental Panel on Climate Change.* Intergovernmental Panel on Climate Change, June 2023. Cambridge, UK: Cambridge University Press. doi.org/10.1017/9781009325844.

Solnit, Rebecca. "Big Oil Coined 'Carbon Footprints' to Blame Us for Their Greed. Keep Them on the Hook." *Guardian*, August 23, 2021. theguardian.com/commentisfree/2021/aug/23/big-oil-coined -carbon-footprints-to-blame-us-for-their-greed-keep-them-on -the-hook.

Supran, Geoffrey, and Naomi Oreskes. "The Forgotten Oil Ads That Told Us Climate Change Was Nothing." *Guardian*, November 18, 2021. theguardian.com/environment/2021/nov/18/the-forgotten-oil -ads-that-told-us-climate-change-was-nothing.

Thacker, Paul D. "In Their Own Words: The Dirty Dozen Documents of Big Oil's Secret Climate Knowledge." *DeSmog*, October 29, 2021. desmog.com/2021/10/29/dirty-dozen-documents-big-oil-secret -climate-knowledge-part-1/.

Union of Concerned Scientists. "The Climate Deception Dossiers." June 29, 2015. ucsusa.org/resources/climate-deception-dossiers.

——. "Smoke, Mirrors, and Hot Air." January 2007. ucsusa.org/sites /default/files/2019-09/exxon_report.pdf.

2: GRIEF

Albrecht, Glenn, Gina M. Sartore, Linda Connor, Nick Higginbotham, Sonia Freeman, Brian Kelly, Helen Stain, et al. "Solastalgia: The Distress Caused by Environmental Change." *Australasian Psychiatry*, February 2007. doi.org/10.1080/10398560701701288.

Almond, Rosamunde, Monique Grooten, Diego Juffe Bignoli, and Tanya Petersen, eds. *Living Planet Report 2022: Building a Nature-Positive Society.* Gland, Switzerland: World Wildlife Fund.

Clayton, Susan, Christie Manning, Kirra Krygsman, and Meighen Speiser. *Mental Health and Our Changing Climate: Impacts, Implications, and Guidance.* American Psychological Association, Climate for Health, and eco America. March 2017. apa.org/news /press/releases/2017/03/mental-health-climate.pdf.

Hickman, Caroline, Elizabeth Marks, Panu Pihkala, Susan Clayton, Eric R. Lewandowski, Elouise E. Mayall, et al. "Climate Anxiety in Children and Young People and Their Beliefs About Government Responses to Climate Change: A Global Survey." *Lancet Planetary Health*, December 2021. doi.org/10.1016/S2542-5196(21)00278-3.

Leiserowitz, Anthony, Edward Maibach, Seth Rosenthal, John Kotcher, Jennifer Carman, Liz Neyens, Jennifer Marlon, et al. *Climate Change in the American Mind, September 2021.* Yale University and George Mason University. New Haven, CT: Yale Program on Climate Change Communication, November 18, 2021. climatecommunication.yale.edu /publications/climate-change-in-the-american-mind-september-2021/.

Leiserowitz, Anthony, Edward Maibach, Seth Rosenthal, John Kotcher, Emily Goddard, Jennifer Carman, Matthew Ballew, Jennifer Marlon, et al. *Climate Change in the American Mind: Beliefs & Attitudes, Fall 2023.* Yale University and George Mason University. New Haven, CT: Yale Program on Climate Change Communication, January 11, 2024. climatecommunication.yale.edu/publications/climate-change -in-the-american-mind-beliefs-attitudes-fall-2023/

Marks, Elizabeth, Caroline Hickman, Panu Pihkala, Susan Clayton, Eric R. Lewandowski, Elouise E. Mayall, Britt Wray, et al. "Young People's Voices on Climate Anxiety, Government Betrayal and Moral Injury: A Global Phenomenon." *Lancet Planetary Health* 5, January 2021. bit.ly/3XvhXuE.

Masson-Delmotte, Valerie, Panmao Zhai, Anna Pirani, Sarah L. Connors, Clotilde Pean, Yang Chen, Leah Goldfarb, et al., eds. *Climate Change 2021: The Physical Science Basis: Working Group I Contribution to the Sixth Assessment Report of the Intergovernmental Panel on Climate Change.* Intergovernmental Panel on Climate Change, June 2023. Cambridge, UK: Cambridge University Press. doi.org/10.1017/9781009157896.

Melley, Brian. "Wildfires Torched Up to a Fifth of All Giant Sequoia Trees." *AP News*, November 19, 2021. apnews.com/article/climate-wildfires-science-environment-and-nature-forests-581747bff7abec7b3d090fb8853d8e1a.

Pihkala, Panu. "Toward a Taxonomy of Climate Emotions." *Frontiers in Climate*, January 14, 2022. doi.org/10.3389/fclim.2021.738154.

Siena College Research Institute. "Think of the Children: The Young and Future Generations Drive US Climate Concern." October 2022. thisisplaneted.org/blog/think-of-the-children-the-young-and-future-generations-drive-u-s-climate-concern.

Sillers, Paul. "Why the Sky Is Still Full of Empty 'Ghost' Flights." *CNN Travel*, April 2, 2022. cnn.com/travel/article/ghost-flights-pandemic-greenpeace-cmd/index.html.

Watts, Nick, W. Neil Adger, Paolo Agnolucci, Jason Blackstock, Peter Byass, Wenjia Cai, Sarah Chaytor, et al. "Health and Climate Change: Policy Responses to Protect Public Health." Lancet Commission on Health and Climate. *Lancet*, November 7, 2015. doi.org/10.1016/S0140-6736(15)60854-6.

3: NATURE

Barrera-Hernández, Laura Fernanda, Mirsha Alicia Sotelo-Castillo, Sonia Beatriz Echeverria-Castro, and Cesar Octavio Tapia-Fonllem. "Connectedness to Nature: Its Impact on Sustainable Behaviors and Happiness in Children." *Frontiers in Psychology*, February 25, 2020. doi.org/10.3389/fpsyg.2020.00276.

Bratman, Gregory N., Christopher B. Anderson, Marc G. Berman, Bobby Cochran, Sjerp De Vries, Jon Flanders, Carl Folke, et al. "Nature and Mental Health: An Ecosystem Service Perspective." *Science Advances*, July 24, 2019. doi.org/10.1126/sciadv.aax0903.

"Children Spend Only Half as Much Time Playing Outside as Their Parents Did." *Guardian*, July 27, 2016. theguardian.com /environment/2016/jul/27/children-spend-only-half-the-time -playing-outside-as-their-parents-did.

Climate Change Resource Center. USDA Climate Change Hubs. July 10, 2022. climatehubs.usda.gov/hubs/climate-change-resource -center.

Dopko, Raelyne L., Colin A. Capaldi, and John M. Zelenski. "The Psychological and Social Benefits of a Nature Experience for Children: A Preliminary Investigation." *Journal of Environmental Psychology*, June 2019. doi.org/10.1016/j.jenvp.2019.05.002.

Edwards-Jones, Andrew, Sue Waite, and Rowena Passy. "Falling into LINE: School Strategies for Overcoming Challenges Associated with Learning in Natural Environments (LINE)." *Education 3–13*, May 6, 2016. doi.org/10.1080/03004279.2016.1176066.

Jia, Bing Bing, Zhou Xin Yang, Gen Xiang Mao, Yuan Dong Lyu, Xiao Lin Wen, Wei Hong Xu, Xiao Ling Lyu, et al. "Health Effect of Forest Bathing Trip on Elderly Patients with Chronic Obstructive Pulmonary Disease." *Biomedical and Environmental Sciences*, March 29, 2016. pubmed.ncbi.nlm.nih.gov/27109132/.

Lee, Kate E., Kathryn J. H. Williams, Leisa D. Sargent, Nicholas S. G. Williams, and Katherine A. Johnson. "40-Second Green Roof Views Sustain Attention: The Role of Micro-Breaks in Attention Restoration." *Journal of Environmental Psychology*, June 2015. doi .org/10.1016/j.jenvp.2015.04.003.

Li, Qing, M. Kobayashi, Y. Wakayama, H. Inagaki, M. Katsumata, Y. Hirata, K. Hirata, et al. "Effect of Phytoncide from Trees on Human Natural Killer Cell Function." *International Journal of Immunopathology and Pharmacology*, October–December 2009. pubmed.ncbi.nlm.nih.gov/20074458/.

Li, Qing, K. Morimoto, M. Kobayashi, H. Inagaki, M. Katsumata, Y. Hirata, K. Hirata, et al. "Visiting a Forest, but Not a City, Increases Human Natural Killer Activity and Expression of Anti-Cancer Proteins." *International Journal of Immunopathology and Pharmacology*, January–March 2008. pubmed.ncbi.nlm.nih .gov/18336737/.

Outdoor Classroom Day. "The Impact of Outdoor Learning and Playtime at School—And Beyond." Project Dirt survey. Outdoor Play and Learning at School, May 2018. outdoorclassroomday.org .uk/resource/the-impact-of-outdoor-learning-and-playtime-at -school-and-beyond/.

Park, Bum Jin, Yuko Tsunetsugu, Tamami Kasetani, Takahide Kagawa, and Yoshifumi Miyazaki. "The Physiological Effects of *Shinrin-Yoku* (Taking in the Forest Atmosphere or Forest Bathing): Evidence from Field Experiments in 24 Forests Across Japan." *Environmental Health and Preventive Medicine*, May 2, 2009. doi.org/10.1007 /s12199-009-0086-9.

Schertz, Kathryn E., and Marc G. Berman. "Understanding Nature and Its Cognitive Benefits." *Current Directions in Psychological Science*, June 24, 2019. doi.org/10.1177/0963721419854100.

Van Hedger, Stephen C., Howard C. Nusbaum, Luke Clohisy, Susanne M. Jaeggi, Martin Buschkuehl, and Marc G. Berman. "Of Cricket Chirps and Car Horns: The Effect of Nature Sounds on Cognitive Performance." *Psychonomic Bulletin and Review*, October 26, 2018. doi.org/10.3758/s13423-018-1539-1.

Zelenski, John M., Raelyne L. Dopko, and Colin A. Capaldi. "Cooperation Is in Our Nature: Nature Exposure May Promote Cooperative and Environmentally Sustainable Behavior." *Journal of Environmental Psychology*, June 2015. doi.org/10.1016/j.jenvp .2015.01.005.

4: ANIMALS

Cassels, Matthew T., Naomi White, Nancy Gee, and Claire Hughes. "One of the Family? Measuring Early Adolescents' Relationships with Pets and Siblings." *Journal of Applied Developmental Psychology* 49, March–April, 2017. Published online January 24, 2017. doi.org/10.1016/j.appdev.2017.01.003.

Hawkins, Roxanne D., and Joanne M. Williams. "Children's Beliefs About Animal Minds (Child-BAM): Associations with Positive and Negative Child–Animal Interactions." *Anthrozoös*, August 17, 2016. doi.org/10.1080/08927936.2016.1189749.

Melson, Gail F. *Why the Wild Things Are: Animals in the Lives of Children.* Cambridge, MA: Harvard University Press, 2021.

Mueller, Megan K. "Is Human–Animal Interaction (HAI) Linked to Positive Youth Development? Initial Answers." *Applied Developmental Science*, January 31, 2014. doi.org/10.1080 /10888691.2014.864205.

Okin, Gregory S. "Environmental Impacts of Food Consumption by Dogs and Cats." *PLoS One*, August 2, 2017. doi.org/10.1371 /journal.pone.0181301.

Sánchez-Bayo, Francisco, and Kris A. G. Wyckhuys. "Worldwide Decline of the Entomofauna: A Review of Its Drivers." *Biological Conservation*, April 2019. doi.org/10.1016/j.biocon.2019.01.020.

5: CURIOSITY

Auteri, Steph. "I Have One Parenting Hack I Swear By, and It's Benign Neglect." *Romper*, October 29, 2021. romper.com/parenting /in-praise-of-benign-neglect.

Bateman, Peter. "Curiosity a Secret Weapon Against Climate Change." United Nations Development Programme, October 31, 2021. undp .org/asia-pacific/news/curiosity-secret-weapon-against-climate-change -story-projek-kitar.

Berger, Warren. *A More Beautiful Question: The Power of Inquiry to Spark Breakthrough Ideas.* New York: Bloomsbury, 2014.

Boudreau, Emily. "A Curious Mind: How Educators and Parents Can Encourage and Guide Children's Natural Curiosity—in the Classroom and at Home." Harvard Graduate School of Education, November 24, 2020. gse.harvard.edu/ideas/usable-knowledge /20/11/curious-mind.

Carrington, Damian. "Elite Minority of Frequent Flyers 'Cause Most of Aviation's Climate Damage.'" *Guardian*, March 31, 2021. theguardian.com/world/2021/mar/31/elite-minority-frequent -flyers-aviation-climate-damage-flights-environmental.

Chouinard, Michelle M. "Children's Questions: A Mechanism for Cognitive Development." *Monographs of the Society for Research in Child Development*, June 28, 2008. doi.org/10.1111/j.1540 -5834.2007.00412.x.

Engel, Susan. "Children's Need to Know: Curiosity in Schools." *Harvard Educational Review*, Winter 2011. doi.org/10.17763 /haer.81.4.h054131316473115.

—. "Many Kids Ask Fewer Questions When They Start School. Here's How We Can Foster Their Curiosity." *Time*, February 23, 2021. time. com/5941608/schools-questions-fostering-curiousity/.

Kahan, Dan M., Asheley Landrum, Katie Carpenter, Laura Helft, and Kathleen Hall Jamieson. "Science Curiosity and Political Information Processing." *Political Psychology*, January 26, 2017. doi.org/10.1111 /pops.12396.

Kim, Hyeji, and Jacob Teter. "Aviation." International Energy Agency Energy System, July 11, 2023. iea.org/energy-system/transport/aviation.

Pappas, Stephanie. "What Do We Really Know About Kids and Screens?" *Monitor on Psychology*, April–May 2020. apa.org /monitor/2020/04/cover-kids-screens.

Parker, Holly. "Tell Me a Story: Why Climate Change Communication Needs to Embrace Our Childlike Curiosity." International Science Council, June 28, 2021. council.science/current/blog/tell-me-a -story-why-climate-change-communication-needs-to-embrace -our-childlike-curiosity/.

Rasmussen, Eric E., Autumn Shafer, Malinda J. Colwell, Shawna
White, Narissra Punyanunt-Carter, Rebecca L. Densley, and Holly
Wright. "Relation Between Active Mediation, Exposure to *Daniel
Tiger's Neighborhood*, and US Preschoolers' Social and Emotional
Development." *Journal of Children and Media*, July 5, 2016. doi.org
/10.1080/17482798.2016.1203806.

6: PROBLEM-SOLVING

Anderson, Jenny. "Why the Danes Encourage Their Kids to Swing
Axes, Play with Fire, and Ride Bikes in Traffic." *Quartz*, October 30,
2018. qz.com/1441424/why-the-danes-encourage-their-kids
-to-play-dangerously.

Barry, Ellen. "In Britain's Playgrounds, 'Bringing in Risk' to Build
Resilience." *New York Times*, March 10, 2018. nytimes.com
/2018/03/10/world/europe/britain-playgrounds-risk.html.

Brooks, Robert, and Sam Goldstein. *Raising Resilient Children:
Fostering Strength, Hope, and Optimism in Your Child.*
Lincolnwood, IL: McGraw Hill Education, 2002.

Kantrowitz, Emma. "How the US Is Embracing Adventure
Playgrounds." Child in the City, April 4, 2017. childinthecity.org
/2017/04/04/how-the-us-is-embracing-adventure-playgrounds/.

King, Barbara J. "Is It Time to Bring Risk Back into Our Kids'
Playgrounds?" National Public Radio, March 15, 2018. npr.org
/sections/13.7/2018/03/15/594017146/is-it-time-to-bring-risk-back
-into-our-kids-playgrounds.

Ockwell-Smith, Sarah. *The Gentle Parenting Book: How to Raise Calmer,
Happier Children from Birth to Seven.* London: Piatkus, 2016.

7: FOOD

Cabrera, Yvette. "The UK Is Winning on Food Waste. Are We?"
National Resources Defense Council, July 30, 2020. nrdc.org/bio
/yvette-cabrera/uk-winning-food-waste-are-we.

"Carbon Footprint Factsheet." Center for Sustainable Systems,
University of Michigan, November 20, 2022. css.umich.edu
/publications/factsheets/sustainability-indicators/carbon-footprint
-factsheet.

Ching, Carrie, Sarah Terry-Cobo, and Arthur Jones. "The Hidden
Costs of Hamburgers." *PBS NewsHour*, August 2, 2012. pbs.org
/newshour/science/the-hidden-costs-of-hamburgers#:~:text
=On%20average%2C%20Americans%20eat%20three,dollars
%20from%20fast%20food%20joints.

Dray, Sally. "In Focus: Food Waste in the UK." House of Lords Library,
UK Parliament, March 12, 2021. lordslibrary.parliament.uk/food
-waste-in-the-uk/.

Food and Agriculture Organization of the United Nations. "The State of World Fisheries and Aquaculture 2020: Sustainability in Action." fao .org/3/ca9229en/ca9229en.pdf.

Natural Resources Defense Council. "Less Beef, Less Carbon: Americans Shrink Their Diet-Related Carbon Footprint by 10 Percent Between 2005 and 2014." March 2017. nrdc.org/sites /default/files/less-beef-less-carbon-ip.pdf.

Poore, Joseph, and Thomas Nemecek. "Reducing Food's Environmental Impacts Through Producers and Consumers." *Science 360*, June 1, 2018. doi.org/10.1126/science.aaq0216.

Ranganathan, Janet, and Richard Waite. "Sustainable Diets: What You Need to Know in 12 Charts." World Resources Institute, April 20, 2016. wri.org/insights/sustainable-diets-what-you -need-know-12-charts.

Ritchie, Hannah. "Food Waste Is Responsible for 6% of Global Greenhouse Gas Emissions." *Our World in Data*, March 18, 2020. ourworldindata.org/food-waste-emissions.

Scarborough, Peter, Paul N. Appleby, Anja Mizdrak, Adam D. M. Briggs, Ruth C. Travis, Kathryn E. Bradbury, and Timothy J. Key. "Dietary Greenhouse Gas Emissions of Meat-Eaters, Fish-Eaters, Vegetarians and Vegans in the UK." *Climatic Change*, June 11, 2014. link.springer.com/article/10.1007%2Fs10584-014-1169-1.

Scarborough, Peter, Michael Clark, Linda Cobiac, Keren Papier, Anika Knuppel, John Lynch, Richard Harrington, et al. "Vegans, Vegetarians, Fish-Eaters and Meat-Eaters in the UK Show Discrepant Environmental Impacts." *Nature Food*, July 20, 2023. doi .org/10.1038/s43016-023-00795-w.

Tubiello, Francesco N., Cynthia Rosenzweig, Giulia Conchedda, Kevin Karl, Johannes Gutschow, Pan Xueyao, Griffiths Obli-Laryea, et al. "Greenhouse Gas Emissions from Food Systems: Building the Evidence Base." *Environmental Research Letters*, June 8, 2021. iopscience.iop.org/article/10.1088/1748-9326/ac018e.

US Environmental Protection Agency. Greenhouse Gas Equivalencies Calculator. epa.gov/energy/greenhouse-gas-equivalencies-calculator.

8: CONSUMERISM

Allen, Davis, Alyssa Johl, Chelsea Linsley, and Naomi Spoelman. "The Fraud of Plastic Recycling." Center for Climate Integrity, February 2024. climateintegrity.org/plastics-fraud.

Bédat, Maxine. "Our Love of Cheap Clothing Has a Hidden Cost—It's Time for a Fashion Revolution." World Economic Forum, April 22, 2016. weforum.org/agenda/2016/04/our-love-of-cheap -clothing-has-a-hidden-cost-it-s-time-the-fashion-industry -changed/.

Beyond Plastics at Bennington College. *The New Coal: Plastics and Climate Change.* October 2021. static1.squarespace.com /static/5eda91260bbb7e7a4bf528d8/t/616ef29221985319611a64e0 /1634661022294/REPORT_The_New-Coal_Plastics_and_Climate -Change_10-21-2021.pdf.

C40, Arup, and University of Leeds. *The Future of Urban Consumption in a 1.5°C World.* June 12, 2019. arup.com/perspectives/publications /research/section/the-future-of-urban-consumption-in-a-1-5c-world.

Center for International Environmental Law. *Plastic and Climate: The Hidden Costs of a Plastic Planet.* May 2019. ciel.org/reports /plastic-health-the-hidden-costs-of-a-plastic-planet-may-2019/.

Geyer, Roland, Jenna R. Jambeck, and Kara Lavender Law. "Production, Use, and Fate of All Plastics Ever Made." *Science Advances*, July 19, 2017. doi.org/10.1126/sciadv.1700782.

Global Plastics Outlook: Economic Drivers, Environmental Impacts and Policy Options. Organisation for Economic Co-operation and Development, February 22, 2022. doi.org/10.1787/de747aef-en.

Koerth, Maggie. "The Era of Easy Recycling May Be Coming to an End." *FiveThirtyEight*, January 10, 2019. fivethirtyeight.com /features/the-era-of-easy-recycling-may-be-coming-to-an-end/.

The Lancet Countdown on Health and Climate Change 2022 Global Launch Event. Lancet YouTube video, October 31, 2022. youtube .com/watch?v=-EbZH4wdOys&t=60s.

Pinsker, Joe. "The Strange Origins of American Birthday Celebrations." *Atlantic*, November 2, 2021. theatlantic.com/family /archive/2021/11/history-birthday-celebrations/620585/.

Quantis. *Measuring Fashion: Environmental Impact of the Global Apparel and Footwear Industries Study*, 2018. quantis.com /wp-content/uploads/2018/03/measuringfashion _globalimpactstudy_full-report_quantis_cwf_2018a.pdf.

United Nations Environment Programme. "How Plastic Is Infiltrating the World's Soils." December 3, 2021. unep.org/news-and-stories /story/how-plastic-infiltrating-worlds-soils.

—. "Our Planet Is Choking on Plastic." December 1, 2022. unep.org/interactives/beat-plastic-pollution/.

—. "What's the Deal with Methane?" October 18, 2022. unep.org /news-and-stories/video/whats-deal-methane.

US Environmental Protection Agency. *National Overview: Facts and Figures on Materials, Wastes and Recycling.* July 1, 2022. epa.gov /facts-and-figures-about-materials-waste-and-recycling/national -overview-facts-and-figures-materials.

World Economic Forum. *The Climate Progress Survey: Business and Consumer Worries and Hopes.* 2021. www3.weforum.org /docs/SAP_WEF_Sustainability_Report.pdf.

9: STEM

Directorate-General for Energy. "The 2021 State of the Energy Union Report." European Commission, November 25, 2021. commission .europa.eu/news/2021-state-energy-union-report-2021-11-25_en.

Generation180. *Brighter Future: A Study on Solar in U.S. K–12 Schools.* September 2020. generation180.org/resource/brighter-future-a-study -on-solar-in-us-schools-2020/.

Grantham Research Institute on Climate Change and the Environment, London School of Economics and Political Science. "How Much Do Renewables Contribute to the UK's Energy Mix and What Policies Support Their Expansion?" London School of Economics and Politican Science, June 23, 2023. lse.ac.uk/granthaminstitute/explainers/how -much-do-renewables-contribute-to-the-uks-energy-mix-and-what -policies-support-their-expansion/.

Heede, Richard. "Tracing Anthropogenic Carbon Dioxide and Methane Emissions to Fossil Fuel and Cement Producers, 1854–2010." *Climatic Change*, November 22, 2013. doi.org/10.1007/s10584 -013-0986-y.

Hernandez-Morales, Aitor. "Renewables Were the EU's Top Source of Power in 2020." *Politico Pro*, October 26, 2021. subscriber. politicopro.com/article/2021/10/renewables-were-the-eus-top -source-of-power-in-2020-3991925.

International Energy Agency. *Net Zero by 2050.* May 2021. iea.org /reports/net-zero-by-2050.

Sierra Club and Climate Parents. *100% Clean Energy School Districts Handbook.* April 2019. sierraclub.org/sites/sierraclub.org /files/Clean-Schools-Handbook.pdf.

Solar Energy Industries Association. "Solar Industry Research Data." seia.org/solar-industry-research-data.

Tabuchi, Hiroko, Brad Plumer, John Schwartz, and Lisa Friedman. "What Are Schools Doing to Go Green?" *New York Times*, September 5, 2018. nytimes.com/2018/09/05/climate/what-are -schools-doing-to-go-green.html.

United Nations. "Secretary-General Warns of Climate Emergency, Calling Intergovernmental Panel's Report 'a File of Shame,' While Saying Leaders 'Are Lying,' Fuelling Flames." April 4, 2022. press.un.org/en/2022/sgsm21228.doc.htm.

US Energy Information Administration. "What Is US Electricity Generation by Energy Source?" 2023. eia.gov/tools/faqs/faq.php?id=427&t=3.

World Health Organization. "More Than 90% of the World's Children Breathe Toxic Air Every Day." October 29, 2018. who.int/news/item/29-10-2018-more-than-90-of-the-worlds-children-breathe-toxic-air-every-day.

10: COMMUNITY

Centola, Damon. *How Behavior Spreads: The Science of Complex Contagions.* Princeton, NJ: Princeton University Press, 2018.

Centola, Damon, Joshua Becker, Devon Brackbill, and Andrea Baronchelli. "Experimental Evidence for Tipping Points in Social Convention." *Science*, June 8, 2018. doi.org/10.1126/science.aas8827.

Devlin, Kat, Moira Fagan, and Aidan Connaughton. "People in Advanced Economies Say Their Society Is More Divided Than Before Pandemic." Pew Research Center, June 23, 2021. pewresearch.org/global/2021/06/23/people-in-advanced-economies-say-their-society-is-more-divided-than-before-pandemic/.

Leiserowitz, Anthony, Edward Maibach, Seth Rosenthal, John Kotcher, Emily Goddard, Jennifer Carman, Marija Verner, et al. *Climate Change in the American Mind: Politics & Policy, Fall 2023*, Yale University and George Mason University. New Haven, CT: Yale Program on Climate Change Communication, November 29, 2023. climatecommunication.yale.edu/publications/climate-change-in-the-american-mind-politics-policy-fall-2023/.

11: ACTIVISM

Ballard, Parissa J., Lindsay T. Hoyt, and Mark C. Pachucki. "Impacts of Adolescent and Young Adult Civic Engagement on Health and Socioeconomic Status in Adulthood." *Child Development*, January 23, 2018. doi.org/10.1111/cdev.12998.

Chan, Wing Yi, Suh-Ruu Ou, and Arthur J. Reynolds. "Adolescent Civic Engagement and Adult Outcomes: An Examination Among Urban Racial Minorities." *Journal of Youth and Adolescence*, May 31, 2014. doi.org/10.1007/s10964-014-0136-5.

Dwyer, Patrick C., Yen-Ping Chang, Jason Hannay, and Sara B. Algoe. "When Does Activism Benefit Well-Being? Evidence from a Longitudinal Study of Clinton Voters in the 2016 U.S. Presidential Election." *PLoS One*, September 5, 2019. doi.org/10.1371/journal.pone.0221754.

Feinberg, Matthew, Robb Willer, and Chloe Kovacheff. "The Activist's Dilemma: Extreme Protest Actions Reduce Popular Support for Social Movements." *Journal of Personality and Social Psychology*, 2020. doi .org/10.1037/pspi0000230.

Gray, Peter. "The Decline of Play and the Rise of Psychopathology in Children and Adolescents." *American Journal of Play*, January 2011. researchgate.net/publication/265449180_The_Decline_of _Play_and_the_Rise_of_Psychopathology_in_Children_and _Adolescents.

Klibert, Jeffrey, Haresh Rochani, Hani Samawi, Kayla Leleux-LaBarge, and Rebecca Ryan. "The Impact of an Integrated Gratitude Intervention on Positive Affect and Coping Resources." *International Journal of Applied Positive Psychology*, April 27, 2019. doi.org /10.1007/s41042-019-00015-6.

CONCLUSION

Hickman, Caroline, Elizabeth Marks, Panu Pihkala, Susan Clayton, R. Eric Lewandowski, Elouise E. Mayall, Britt Wray, et al. "Climate Anxiety in Children and Young People and Their Beliefs About Government Responses to Climate Change: A Global Survey." *Lancet Planetary Health*, December 2021. doi.org/10.1016/S2542 -5196(21)00278-3.

Sillers, Paul. "Why the Sky Is Still Full of Empty 'Ghost' Flights." CNN Travel, April 2, 2022. cnn.com/travel/article/ghost-flights -pandemic-greenpeace-cmd/index.html.

United Nations Children's Fund. *The Climate Crisis Is a Child Rights Crisis: Introducing the Children's Climate Risk Index.* August 2021. unicef.org/reports/climate-crisis-child-rights-crisis.

Watts, Nick, W. Neil Adger, Paolo Agnolucci, Jason Blackstock, Peter Byass, Wenjia Cai, Sarah Chaytor, et al. "Health and Climate Change: Policy Responses to Protect Public Health." Lancet Commission on Health and Climate. *Lancet*, November 7, 2015. doi.org/10.1016 /S0140-6736(15)60854-6. Epub June 25, 2015. PMID: 26111439.

Acknowledgments

To the parents, scientists, teachers, and climate leaders who shared their hopes and fears, patiently answered my many, many questions, and lent their expertise. Your resilience is revolutionary. To Maisie Tivnan and the team at Workman Publishing, for taking on this project and making it happen. Without your support and incisive edits, this book would not exist. To my agent, Emma Bal, and the team at the Madeleine Milburn Literary Agency, for the opportunity of their mentorship program and reminding me to trust the process. To Elissa, B. A., Tanya, Jessie, Jason, Molly, Amy, Karly, and the Glue Factory writing group for the early draft edits, the coffee, and pulling me away from the computer enough to get out of my head. (E., I do hear your journalistic voice in my brain.) To all the dogs for whom I've had the privilege of being one of the humans, for the emotional support and teachings about life and loss. To Mom and Dad, for showing me the beauty and fragility of the natural world from a young age, always encouraging me even when you didn't really understand my job, and for the countless hours of childcare. To my sister, for always checking in, often with photos of a gigantic goldendoodle. To my brother, for his thoughts along the way and always being up for a road trip. To my child, for reminding me what we fight for every day. To this incredible planet that we call home, that inspired this book. Thank you.

About the Author

BRIDGET SHIRVELL is a parent and writer. Her food, parenting, and environmental reporting has appeared in publications including the *New York Times*, the *Washington Post*, *Martha Stewart Living*, *Good Housekeeping*, and more. She's been an environmentalist for as long as she can remember; her climate solutions work took on new urgency after becoming a parent. Bridget lives in an old house in Mystic, Connecticut, with her daughter and dog. She can be found online at breeshirvell.com and on social media at @breeshirvell.

Index

obsessiveness, 35
overwhelmed feelings, 35
ozone, 6

packaging, food, 114–116
palm oil, 110
parasympathetic nervous
    system, 55
parents, talking to, 22–25
Parents for Future, 22, 23,
    24–25, 164, 169
personal carbon footprint,
    use of term, 11
pets, 63–75
physical symptoms of
    climate anxiety, 35
Plant Hardiness Zones, 4
plant-based diets, 106, 109,
    125. *See also* food
planting things, 174–175
plants, 59–60. *See also*
    gardening; nature; trees
plastics, 128, 131–132, 135
play
    emotion and, 99, 102
    free, 179–180
    unstructured, 97–98
play deficit, 179
playgrounds, 96–97
pollinator habitats, 155
powerlessness, 34–35
privilege, 20–21
problem-solving, 93–104

property destruction,
    181–182

rain barrels, 154
reading/books, 85
ready-to-eat items, limiting,
    109–110
recycling programs,
    problems with, 132
regenerative travel, 87–88
renewable energy, 146–148,
    149–152, 171–174
resilience
    activism and, 167
    community and, 159,
        164–165
    curiosity and, 80
    description of, ix
    problem-solving and,
        95–99
    risk and, 97–98
respiratory illnesses, 54
risk, playgrounds and,
    96–98
routines, changing, 86

schools, 48–50, 61, 71–72,
    139, 171–174
Science Moms, 11–12, 22,
    23, 47, 169, 170–171, 175
screen time, 84
seafood, 108
seasons, time outside in, 48